INFRARED SPECTRAL INTERPRETATION
A Systematic Approach

INFRARED SPECTRAL INTERPRETATION
A Systematic Approach

Brian Smith

CRC Press
Boca Raton   London   New York   Washington, D.C.

Library of Congress Cataloging-in-Publication Data

Smith, Brain C.
    Infrared spectral interpretation : a systematic approach / Brain C. Smith.
        p.   cm.
    Includes bibliographical references and index.
    ISBN 0-8493-2463-7 (alk. paper)
    1. Infrared spectroscopy. I. Title.
QD96.I5S583 1998
543′.0858—dc21   98-37190
                                                                                                                                                                      CIP

    This book contains information obtained from authentic and highly regarded sources. Reprinted material is quoted with permission, and sources are indicated. A wide variety of references are listed. Reasonable efforts have been made to publish reliable data and information, but the author and the publisher cannot assume responsibility for the validity of all materials or for the consequences of their use.

    Neither this book nor any part may be reproduced or transmitted in any form or by any means, electronic or mechanical, including photocopying, microfilming, and recording, or by any information storage or retrieval system, without prior permission in writing from the publisher.

    The consent of CRC Press LLC does not extend to copying for general distribution, for promotion, for creating new works, or for resale. Specific permission must be obtained in writing from CRC Press LLC for such copying.

    Direct all inquiries to CRC Press LLC, 2000 Corporate Blvd., N.W., Boca Raton, Florida 33431.

    Trademark Notice: Product or corporate names may be trademarks or registered trademarks, and are only used for identification and explanation, without intent to infringe.

© 1999 by CRC Press LLC

No claim to original U.S. Government works
International Standard Book Number 0-8493-2463-7
Library of Congress Card Number 98-37190
Printed in the United States of America 1 2 3 4 5 6 7 8 9 0
Printed on acid-free paper

Preface

Infrared spectroscopy is an important field of chemical analysis. Anyone interested in identifying the components or measuring the concentrations of molecules in a sample should become familiar with this technique. The first task in the infrared analysis of any sample is obtaining the spectrum itself. This involves preparing the sample, placing it in an infrared spectrometer, and measuring the response of the sample to infrared light. I refer you to my book, *Fundamentals of Fourier Transform Infrared Spectroscopy*, also published by CRC Press, for instruction in these topics. However, obtaining a spectrum is only half the battle. The second half is figuring out what the spectrum means, and this volume is designed to assist you in that task.

Many excellent books on infrared interpretation have been published over the last 50 years, so why yet another book on the subject? What this volume offers is a new approach to interpretation, a new way of looking at infrared spectra, and a new way of teaching the art and science of this field. Interpreting spectra is not about memorizing peak positions; it is an exercise in *pattern recognition*. All of the aspects of a spectrum, the position, height, width, and *pattern* of bands carry important information about a sample. Interpreting spectra is also a process. By asking the right questions of a spectrum in the right order, more information will be obtained than by randomly identifying specific peaks. This book emphasizes the use of all the information in a spectrum to improve the quality of interpretations, and includes a 10-step program readers should follow when analyzing their spectra to insure the process is systematic. The organization of this book, with clearly defined terms, numerous example spectra, many molecular diagrams, and a large number of fully solved problem spectra, also represents a contribution to this field. The logic of the approach taken in this book has been born out by my experience teaching an infrared spectral interpretation course for Spectros Associates. Thousands of attendees and thousands of positive comments have convinced me that my approach is sound.

The intended audience for this book is anyone interested in analyzing samples using infrared spectroscopy. However, it is written at a basic level and assumes little on the part of the reader. A familiarity with the terms and symbols of organic chemistry would be of help in reading this book, but is by no means essential. All terms that appear in *italics* in the text are defined in the glossary at the end of the book. There is a minimum of mathematics in the text, all chemical structures are clearly drawn, and all chemical terms are clearly defined. Any person with a technical background should be able to pick up this book and learn something from it. However, experienced practitioners of infrared interpretation will find this book a useful reference. The summarization of group wavenumbers in Appendix II and the firm theoretical foundation of infrared absorbance given in Chapter 1 should be of particular use to advanced readers. I have included >30 unknown spectra that can serve as exercises for the reader. The surest way of learning infrared interpretation is to work through these exercises. The solution to these problems with the correct interpretation process fully described is given in Appendix I. The exercises make the book useful as a textbook in courses on instrumental analysis, organic chemistry, or the spectrometric identification of organic compounds.

The subtitle of the book, *A Systematic Approach*, is taken seriously. Chapter 1 contains a discussion of the science of infrared interpretation. The nature of light and molecular vibrations is introduced, and then a discussion of how and why molecules absorb infrared radiation is given. Next, the meaning and importance of peak heights, intensities, and widths is emphasized. The first chapter ends with a 10-step systematic process on how to interpret spectra. Chapters 2 through 8 are organized by functional group and bear on the art of infrared interpretation. There are chapters on hydrocarbons, C-O bonds, carbonyl groups, C-N bonds, polymers, inorganics, and organic molecules containing sulfur, silicon, and halogens. Each functional group chapter is introduced with a discussion of the nomenclature and structure of that specific group. You need not know what a methyl group is, or be able to tell the difference between a primary and tertiary amine to read these chapters (although you will know about methyl groups and distinguishing amines after reading these chapters). Everything is explained to you in the text. After familiarizing you with the structures and names of functional groups, each of these chapters continues with the important diagnostic infrared bands for that functional group. Diagrams will be used to illustrate molecular vibrations. Each chapter contains many example spectra. These are laid out in "landscape" orientation (parallel to the long axis of the page) so the spectra can be made as large as possible for easy viewing. The important peaks in each spectrum are clearly marked, and accurate wavenumber tables showing the position and assignment of important peaks, are found below each spectrum. The >50 example spectra in this book go a long way toward giving readers a feeling for the patterns and subtleties of the spectra of different functional groups.

Chapter 9 is a discussion of interpretation aids. Topics covered include spectral atlases, subtraction, library searching, "expert" software, and IR interpretation resources available on the Internet. The point of Chapter 9 is to make the reader aware of the many tools that are available to make interpretation faster and easier. However, it must be emphasized that these tools are not a replacement for knowledge of infrared interpretation, and that they are easier to use if one has an understanding of infrared interpretation to begin with. The answers to each unknown spectrum are listed in Appendix 1. This is more than a simple listing of the correct peak assignments. The name and chemical structure of each molecule are given, along with a detailed description of the proper interpretation process for each spectrum. These detailed descriptions lie at the heart and soul of what I am trying to teach in this book. Appendix II contains a listing of all the wavenumber tables throughout the entire book. This collection of tables is an excellent summary of all the knowledge contained herein. I encourage you to consult these tables often when interpreting unknown spectra. The glossary that ends the book contains concise definitions of many of the terms that appear in italics in the body of the book.

The spectra in this book came from many sources, including commercially available spectral libraries, and spectra downloaded from the Internet. All spectra were recorded with modern FTIR instruments. Only high quality, well resolved, high signal-to-noise ratio spectra of pure materials were chosen for inclusion in this book. The wavenumber axis of FTIRs is inherently stable and precise thanks to the use of on-board He-Ne lasers as internal wavenumber standards. Thus, the accuracy of peak positions recorded in this book is as good as modern FTIR technology allows.

A note on the use of units in this book. Infrared interpretation textbooks commonly use the terms "frequency" and "group frequency" when referring to the peak positions of infrared bands. The use of the term "frequency" in this context is a misnomer. The units of frequency are Hertz, which are measured in sec^{-1}. Infrared spectra are usually plotted with the X-axis in units of wavelength or wavenumber, which have units of length or $length^{-1}$. To properly state the position of an infrared band, the terms "wavenumber" and "group wavenumber" should be used. Thus, to describe the C=O stretching band of acetone, one need simply say, "acetone absorbs at 1715 cm^{-1}." The terms wavenumber and group wavenumber will be used throughout this book.

I take full responsibility for all errors and omissions in this book. Please address all comments and criticisms of the book to me at the e-mail address below. Happy interpreting!

Brian C. Smith
Shrewsbury, MA
bcsmith@ma.ultranet.com
October, 1998

Acknowledgments

In the undertaking of any project, there are many people who must be thanked for their contributions, large and small, to the endeavor. The writing of this book is no exception. The manuscript was carefully proofread by Laurie Sparks of Westinghouse WIPP. Having also read the manuscript of my first book, Laurie has become my unofficial proofreader. I cannot thank her enough for the time she has put into making my volumes more readable. The manuscript was also perused by Professor Terry Morrill of Rochester Institute of Technology. Dr. Morrill is a well-known expert in the spectral identification of organic compounds and co-authored one of the first books in this field. I was fortunate to have Dr. Morrill teach my first course in IR spectral interpretation as an undergraduate at RIT. Having him review the manuscript is a thrill, since my former teacher is now my colleague.

Books of this nature are impossible to write without the use of many infrared spectra as examples. The Sadtler Division of Bio-Rad was kind enough to allow me to use many of their spectra in this book. Greg Angsten at Photometrics and Ed Bartick at the FBI were kind enough to grant me permission to use some of their spectra available off the Internet. This endeavor would be impossible without their kindness.

Many of the spectra in this book were displayed and plotted using GRAMS software from Galactic Industries of Salem, NH. In addition the spectra I obtained from the Internet were found via links at the Galactic Industries website, www.galactic.com. My main source of technical help at Galactic Industries, Mr. Anthony Nip, was always available to answer my questions, and was cheerful about it too. Thank you, Tony!

Most of the molecular structures and drawings of molecular vibrations were created using the ChemWindow software package from SoftShell division of Bio-Rad. I want to thank SoftShell for supplying a copy of the software and Rebecca Roseberry for being extremely helpful when I had questions.

This book has evolved over the last 6 years, as it has been used as the main text for the infrared interpretation course I teach for Spectros Associates. Literally thousands of people have taken the course and made thousands of comments on how to improve the book. I would like to thank all of my course attendees for their constructive criticism.

Last, I would like to thank my editors at CRC Press, Felicia Shapiro, and Mimi Williams for their hard work and common sense. Felicia learned early on that I am a person that responds well to deadlines...and it is possible that this book would never have been finished without Felicia's friendly but firm guidance. Mimi was helpful in insuring my camera-ready copy really was "camera-ready." Thanks a lot, ladies!

In conclusion, and for safety's sake, I would like to thank the entire Eastern and Western Hemispheres.

To the lovely ladies in my life...

Marian, Eleanor, and Isabel

Table of Contents

Chapter 1 The Basics of Infrared Interpretation — 1
I. Introduction
A. The Advantages of Infrared Spectroscopy — 1
B. The Disadvantages of Infrared Spectroscopy — 2

II. The Basics of Infrared Absorbance
A. The Properties of Light — 2
B. What is an Infrared Spectrum? — 5
C. The Nature of Molecular Vibrations: Normal Modes — 7

III. How Molecules Absorb Infrared Radiation
A. The First Necessary Condition for Infrared Absorption — 7
B. The Second Necessary Condition for Infrared Absorption — 12

IV. The Origin of Infrared Peak Positions, Intensities, and Widths
A. Peak Positions — 15
   1. The Harmonic Oscillator Model of Molecular Vibrations — 15
   2. Real World Molecules: Anharmonic Vibrations — 17
B. The Origin of Peak Intensities — 18
C. The Origin of Peak Widths — 19
D. The Origin of Group Wavenumbers — 24

V. Infrared Spectral Interpretation: A Systematic Approach
A. Dealing With Mixtures — 25
B. Properly Performing Identities — 26
C. Infrared Spectral Interpretation: A 10-Step Approach — 27

VI. Conclusion — 29
Bibliography — 29

Chapter 2 Hydrocarbons — 31
I. Alkanes
A. Structure and Nomenclature — 31
B. The Infrared Spectra of Straight Chain Alkanes
   1. C-H Stretching Vibrations — 32
   2. C-H Bending Vibrations — 35
   3. Estimating Hydrocarbon Chain Length from Infrared Spectra — 36
C. The Infrared Spectra of Branched Alkanes
   1. Structure and Nomenclature — 38
   2. Spectra of Branched Chain Alkanes — 40

II. Alkenes
A. Structure and Nomenclature 43
B. The Infrared Spectra of Alkenes
    1. C-H Stretching 43
    2. C=C Stretching 45
    3. C-H Bending 45

III. Alkynes
A. Structure and Nomenclature 47
B. Infrared Spectra of Alkynes 47

IV. Aromatic Hydrocarbons
A. Structure and Nomenclature 49
B. The Infrared Spectra of Aromatic Molecules
    1. Overview: The Infrared Spectrum of Benzene 50
    2. The Infrared Spectra of Mono- and Disubstituted Benzene Rings 52
    3. Methyl Groups Attached to Benzene Rings 54
Bibliography 58
Problem Spectra 58

Chapter 3 Functional Groups Containing the C-O Bond 67

I. Alcohols and Phenols
A. Structure and Nomenclature 67
B. The Infrared Spectra of Alcohols: The Effects of Hydrogen Bonding 68
C. The Infrared Spectra of Saturated Alcohols 70
D. The Infrared Spectra of Aromatic Alcohols: Phenols 72
E. Distinguishing Alcohols from Water 75

II. Ethers
A. Structure and Nomenclature 75
B. The Infrared Spectra of Ethers 76
C. Saturated Ethers 76
D. The Infrared Spectra of Aromatic Ethers 78
E. CH_3 and CH_2 Groups Attached to an Oxygen 78
Bibliography 81
Problem Spectra 81

Chapter 4 The Carbonyl Functional Group
I. Introduction 91

II. Ketones
A. Structure and Nomenclature 92
B. The Infrared Spectra of Ketones 93

III. Aldehydes
A. Structure and Nomenclature 96
B. The Infrared Spectroscopy of Aldehydes 96

IV. Carboxylic Acids and Their Derivatives
A. Carboxylic Acids 98
   1. Structure and Nomenclature 98
   2. The Infrared Spectra of Carboxylic Acids 97
B. Carboxylic Acid Salts: Carboxylates
   1. Structure and Nomenclature 103
   2. The Infrared Spectra of Carboxylates 103
C. Acid Anhydrides
   1. Structure and Nomenclature 105
   2. Infrared Spectroscopy of Acid Anhydrides 105

V. Esters
A. Structure and Nomenclature 108
B. The Infrared Spectra of Esters: The Rule of Three 108
C. Saturated Esters 109
D. Aromatic Esters 112

VI. Organic Carbonates
A. Structure and Nomenclature 112
B. The Infrared Spectra of Organic Carbonates 114

VII. Summary 114
Bibliography 115
Problem Spectra 115

Chapter 5 Organic Nitrogen Compounds
I. Introduction 125

II. Amides
A. Structure and Nomenclature 125
B. The Infrared Spectra of Primary, Secondary, and Tertiary Amides
   1. Primary Amides 128
   2. Secondary Amides 128
   3. Tertiary Amides 130
   4. Proteins 132

III. Imides
A. Structure and Nomenclature 132
B. The Spectroscopy of Imides 133

IV. Amines
A. Structure and Nomenclature 135
B. The Infrared Spectra of Primary, Secondary, and Tertiary Amines
    1. Primary Amines 136
    2. Secondary Amines 138
    3. Tertiary Amines 140
    4. Methyl Groups Attached to an Amine Nitrogen 140

V. Nitriles: The C≡N Bond
A. Structure and Nomenclature 141
B. Spectroscopy of the C≡N Bond 141

VI. The Nitro Group
A. Structure and Nomenclature 143
B. The Spectroscopy of the NO_2 Group 144
Bibliography 146
Problem Spectra 146

Chapter 6 Organic Compounds Containing Sulfur, Silicon, and Halogens

I. Organic Sulfur Compounds 153
A. Thiols (Mercaptans)
    1. Structure and Nomenclature 153
    2. Infrared Spectra of Thiols 155
B. Molecules Containing Sulfur/Oxygen Bonds
    1. Structure and Nomenclature 155
    2. Spectra of Compounds Containing One S=O Bond 156
    3. Spectra of Compounds with Two S=O Bonds 158

II. Organic Silicon Compounds
A. Structure and Nomenclature of Siloxanes (Silicones) 158
B. Spectra of Siloxanes 160

III. Halogenated Organic Compounds 160
Bibliography 163

Chapter 7 Inorganic Compounds
I.   Introduction   165

II. Inorganic Sulfates   168

III. Silica   168

IV. Inorganic Carbonates   171

V.   Nitrates   171

VI. Phosphates   173
Bibliography   175

Chapter 8 Infrared Spectra of Polymers
I. Introduction   177

II. Recyclable Plastics   175
A. Low Density and High Density Polyethylene   177
B. Polypropylene   180
C. Polystyrene   180
D. Polyethylene Terephthalate   184

III. Engineering Plastics
A. Polyamides   184
B. Acrylates   184
C. Diisocyanates and Polyurethanes   187
D. Polycarbonates   190
E. Polyimides   190
F. Polytetrafluoroethylene   190
Bibliography   194

Chapter 9 Spectral Interpretation Aids
I. Spectral Atlases   195

II. Spectral Subtraction   196
A. Optimizing the Subtraction Factor   196
B. Subtraction Artifacts   198

III. Spectral Library Searching   199
A. Search Algorithms   201
B. Interpreting Search Results   201

C. Subtract and Search Again	202
IV. Infrared Interpretation Software Packages	204
V. Infrared Interpretation and the Internet	204
Bibliography	205

Appendix I Answers to Problem Spectra   207

Appendix II Group Wavenumber Tables   243

Glossary   251

Index   259

Chapter 1

The Basics of Infrared Interpretation

I. Introduction

Infrared spectroscopy is the study of the interaction of infrared light with matter. The fundamental measurement obtained in infrared spectroscopy is an *infrared spectrum*, which is a plot of measured infrared intensity versus wavelength (or wavenumber) of light. An instrument used to obtain an infrared spectrum is called an *infrared spectrometer*. There are several kinds of spectrometers in the world used to obtain infrared spectra. The most prevalent type of spectrometer is called a *Fourier Transform Infrared Spectrometer* (FTIR). For a detailed discussion of instrumentation, sample preparation, and quantitative analysis, please see the author's book on FTIR [1].

Infrared spectroscopy is sensitive to the presence of chemical *functional groups* in a sample. A functional group is a structural fragment within a molecule. Functional groups often have chemical properties that are the same from molecule to molecule. For example, the C=O bond of a ketone or the CH_3 group in a hydrocarbon are examples of functional groups.

The most powerful aspect of infrared spectroscopy is that it allows you to identify unknowns. As you will see later in this book, once the wavenumber positions of the bands of a functional group are known, this information can be used to identify that functional group in many samples. A second use of infrared spectra is in confirming *identities*. Identities involve comparing the spectra of two samples to each other to determine whether the samples have the same composition. Finally, the peak intensities in an infrared spectrum are proportional to concentration, so infrared spectra can be used to measure concentrations as well. This book will teach you how to identify unknowns and how to confirm identities using infrared spectra.

A. The Advantages of Infrared Spectroscopy

It is important to know the strengths and weaknesses of any analytical technique, so the technique may be used in the proper way. Infrared spectroscopy works well on some samples, and poorly on others. The purpose of this section is to discuss the good and bad points of infrared spectroscopy so you can use the technique appropriately.

Infrared spectroscopy has many advantages as a chemical analysis technique. First, it is a universal technique. Solids, liquids, gases, semi-solids, powders, and polymers are all routinely analyzed. Second, infrared spectra are information rich; the peak positions, intensities, widths, and shapes in a spectrum all provide useful information. Third, infrared spectroscopy is a relatively fast and easy technique. The majority of samples can be prepared, scanned, and the results plotted in less than five minutes. Infrared spectroscopy is also a sensitive technique. By interfacing an infrared spectrometer to a gas chromatograph, the infrared spectrum of as little as 5 nanograms (5×10^{-9} gram) of material can be obtained. Micrograms (10^{-6} gram) of material can be detected routinely. Finally, infrared instruments are relatively inexpensive. As of this writing, a quality infrared instrument, software, computer, and sampling accessories can be purchased for under $20,000. This is inexpensive for a laboratory instrument.

B. The Disadvantages of Infrared Spectroscopy

Despite the many advantages of infrared spectroscopy, it is not well suited for certain samples. Since infrared spectroscopy is sensitive to the presence of functional groups in a sample, a sample must contain chemical bonds to have an infrared spectrum. Single atomic entities contain no chemical bonds, and hence do not absorb infrared radiation. Thus, atoms or monatomic ions do not have infrared spectra. The noble gases such as helium and argon do not have infrared spectra because they exist as individual atoms. Monoatomic ions such as Pb^{+2} dissolved in water are not chemically bonded to anything and do not have an infrared spectrum. Analyzing drinking water for the presence of lead would be an inappropriate use of infrared spectroscopy. In general, it is impossible for an infrared spectrometer to measure the level of an element in a substance, unless the element is present as part of a molecule whose spectrum can be detected.

Another class of substances that do not absorb infrared radiation are homonuclear diatomic molecules. These molecules consist of two identical atoms, such as $N \equiv N$ and $O=O$. Homonuclear diatomic molecules do not possess infrared spectra due to their symmetry (see later in this chapter). It would be impossible to determine the amount of oxygen or nitrogen in ambient air using an infrared spectrometer.

Another limitation of infrared spectroscopy is its use in analyzing complex mixtures. Infrared spectroscopy works best on pure substances, since all bands can be assigned to a single molecular structure. If a sample's composition is complex, its spectrum will be complex and it will be hard to know which infrared bands are due to which molecules. The best way to deal with complex mixtures is to simplify the composition of the sample by purifying it before taking its spectrum. Other ways of dealing with mixture spectra will be discussed later in this chapter.

Aqueous solutions are also difficult to analyze using infrared spectroscopy. Water is a strong infrared absorber (see Figure 1.13), and dissolves many of the infrared transparent materials used in cells and windows, such as NaCl and KBr. There are sampling techniques that allow the infrared spectra of aqueous solutions to be measured, but these techniques are not very sensitive. In general, infrared spectroscopy is not used to analyze trace amounts of material dissolved in water.

In conclusion, infrared spectroscopy is used to analyze a wide variety of samples, but it cannot solve every chemical analysis problem. If you are familiar with the advantages and limitations of this technique, you will minimize the time spent analyzing inappropriate samples.

II. The Basics of Infrared Absorbance

A. The Properties of Light

Light can be thought of as having wave-like and particle-like properties. For now, we will consider light a wave. Light beams are composed of electric and magnetic waves. These waves oscillate in planes perpendicular to each other, and the light wave traverses through space in a direction perpendicular to the planes containing the electric and magnetic waves. Light is called *electromagnetic radiation* because it has electric and magnetic parts. The electric part of light called the *electric vector* interacts with molecules to cause infrared absorbance. The amplitude of the electric vector changes over time and has the form of a sine wave, as shown in Figure 1.1.

Different types of light are denoted by their *wavelength*. A wavelength is the distance between adjacent crests or troughs of a wave, as seen in Figure 1.1. Wavelength is denoted by the small Greek letter lambda (λ). Different types of light have different wavelengths. For example, infrared radiation is longer in wavelength than visible light, but shorter than microwaves and radio waves. This is illustrated in Figure 1.2.

Another way of denoting different types of light is by the light wave's *frequency*. A light wave's frequency is denoted by the Greek letter nu (ν). The frequency of a light wave equals the number of cycles the wave undergoes in a second. A cycle is considered complete when a light wave crosses the X-axis twice. The wave in Figure 1.1 undergoes almost three cycles. Frequency is measured in *Hertz* (Hz) whose units are sec^{-1}.

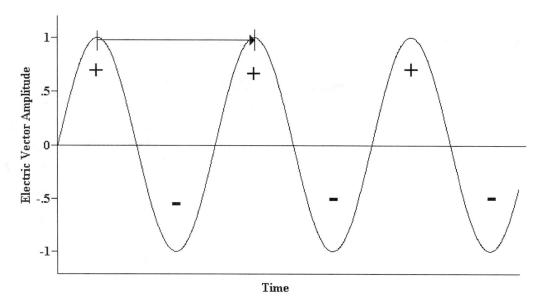

Figure 1.1. A plot of the amplitude of the electric vector part of a light wave versus time. The arrow denotes the distance between adjacent crests, and is called the wavelength, λ.

The frequency, wavelength, and speed of light (c) are related to each other via the following equation:

$$c = \nu\lambda \tag{1.1}$$

where
  c = the speed of light (3×10^{10} cm/second)
  ν = frequency in Hertz (sec^{-1})
  λ = wavelength in cm

This equation shows that the product of the frequency and wavelength of a light wave is equal to a constant, the speed of light.

In infrared spectroscopy, units called *wavenumbers* are normally used to denote different types of light. A wavenumber is defined as the reciprocal of the wavelength as follows:

$$W = 1/\lambda \tag{1.2}$$

where
  W = wavenumber in cm^{-1}
  λ = wavelength in cm

If λ is measured in cm, then W is reported as cm^{-1}, sometimes called reciprocal centimeters. W is a measure of the number of cycles in a light wave per centimeter. If we substitute Equation 1.2 into 1.1 and solve for c, we obtain

4 The Basics of Infrared Interpretation

or
$$c = \nu/W \tag{1.3}$$
$$\nu = cW$$

Equations 1.1 through 1.3 show that light waves may be described by their frequency, wavelength, or wavenumber. These equations also show that these three quantities are related to each other. Throughout this book, we will typically refer to light waves by their wavenumber. However, at times it will be more convenient to refer to a light wave's frequency or wavelength. The wavenumbers of several different types of light are shown in Figure 1.2.

> 14,000 cm^{-1} Visible UV & X-rays	14,000 to 4000 cm^{-1} Near Infrared	4000 to 400 cm^{-1} Mid Infrared	400 to 4 cm^{-1} Far Infrared	< 4 cm^{-1} Microwaves Radio Waves

◄―――――――――――――――――――――――――――――――――►

Higher Wavenumber	Lower Wavenumber
Higher Frequency	Lower Frequency
Higher Energy	Lower Energy
Shorter Wavelength	Longer Wavelength

Figure 1.2. The electromagnetic spectrum, showing the wavenumber ranges for different types of light.

As can be seen in Figure 1.2, microwaves and radio waves occur at low wavenumber. *Mid-infrared radiation* is found between 4000 and 400 cm^{-1}. At higher wavenumbers visible light, ultraviolet light and X-rays are found. This book will focus exclusively upon the interaction of mid-infrared light with matter. The vast majority of molecules in the universe absorb mid-infrared light, which is why mid-infrared spectroscopy is so useful.

As mentioned above, light also has particle-like properties. A particle of light is called a *photon*. A photon has no mass, but it does have energy. The frequency of a photon (E) is directly proportional to its energy as follows:

$$E = h\nu \tag{1.4a}$$

If we substitute Equation 1.3 into 1.4a, we obtain

$$E = hcW \tag{1.4b}$$

where
  E = photon energy in joules
  h = planck's constant (6.63 x 10^{-34} Joule-second)
  ν = frequency in Hertz (sec^{-1})
  c = the speed of light (3x10^{10} cm/second)
  W = wavenumber in cm^{-1}

According to these equations, high wavenumber light has more energy than low wavenumber light.

B. What Is an Infrared Spectrum?

Another name for infrared radiation is heat. All objects in the universe at a temperature above absolute zero give off infrared radiation. Your body (and this book) is giving off infrared radiation even as you read. When infrared radiation interacts with matter, it can be absorbed, causing the chemical bonds in the material to vibrate. The presence of chemical bonds in a material is a necessary condition for infrared absorbance to occur. Functional groups tend to absorb infrared radiation in the same wavenumber range regardless of the structure of the rest of the molecule. For instance, the C=O stretch of a carbonyl group occurs at ~ 1700 cm^{-1} in many different types of molecules. This means there is a correlation between the wavenumbers at which a molecule absorbs infrared radiation and its structure. This correlation allows the structure of an unknown molecule to be determined from its infrared spectrum, and allows spectra of different samples to be compared to each other to see if they are the same. The correlation between infrared band positions and chemical structure is what makes infrared spectroscopy a useful chemical analysis tool.

A plot of measured infrared radiation intensity versus wavenumber is known as an infrared spectrum. An example of an infrared spectrum, the spectrum of polystyrene, is shown in Figure 1.3.

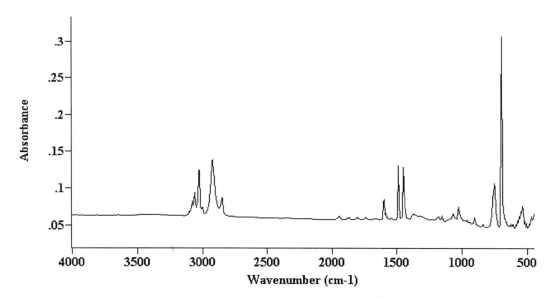

Figure 1.3. The infrared spectrum of polystyrene plotted in absorbance units on the Y-axis.

Most modern infrared spectra are plotted with wavenumber on the X-axis, but some older spectra are plotted in wavelength. Infrared spectra are conventionally plotted with high wavenumber on the left and low wavenumber on the right. This means that as a spectrum is read from left to right, one is reading from high energy to low energy. Most commonly, spectra are obtained from 4000 to 400 cm^{-1}. The wavenumber range of a spectrum is usually determined by the infrared spectrometer on which the sample is examined.

Note that in Figure 1.3 that the Y-axis is plotted in absorbance, which is defined as follows:

$$A = \log(I_0/I) \qquad (1.5)$$

where

A = absorbance
I = light intensity with a sample in the infrared beam (sample spectrum)
I_0 = light intensity measured with no sample in the infrared beam (background spectrum)

6 The Basics of Infrared Interpretation

The I_0 in Equation 1.5 is the "background" spectrum typically measured before the sample spectrum on an FTIR. The purpose of I_0 is to measure the contribution of the spectrometer and the environment to a spectrum. The parameter I contains contributions from the sample, instrument, and environment. By taking the ratio of I_0 to I, the instrument and environment contributions cancel and only the sample's spectrum is retained. Absorbance is the log of I_0/I, and is a unitless quantity. The upward pointing peaks in an absorbance spectrum represent wavenumbers at which the sample absorbed infrared radiation.

The Y-axis of an infrared spectrum can also be plotted in *transmittance*. Transmittance is defined as follows:

$$T = I/I_0 \qquad (1.6)$$

where
  T = transmittance
  I = light intensity with a sample in the infrared beam
  I_0 = light intensity with no sample in the infrared beam

In a transmittance spectrum, the peaks point down and represent wavenumbers at which the sample transmitted little infrared radiation. The units of transmittance are percent, with the scale typically from 0 to 100%. The transmittance spectrum of polystyrene is seen in Figure 1.4.

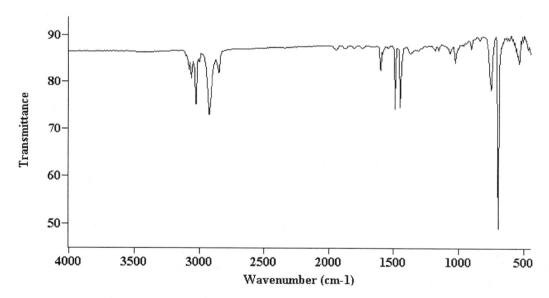

Figure 1.4. The transmittance spectrum of polystyrene.

If we combine Equations 1.5 and 1.6, we obtain a relationship between absorbance and transmittance

$$A = \log(1/T) \qquad (1.7)$$

There is some debate in the infrared spectroscopy world about whether spectra should be plotted in absorbance or transmittance. For spectra being used in quantitative analysis (measuring concentrations), the Y axis units <u>must</u> be in absorbance. This is because *Beer's Law* states that absorbance and concentration are linearly proportional (for more on Beer's Law, see below). Transmittance and concentration are not linearly proportional, making transmittance spectra ill-suited for quantitative analysis. When interpreting a spectrum for qualitative purposes one is not concerned about concentrations, but the positions, relative intensities, and widths of the features in a spectrum. In these instances, it is a matter of personal preference whether to plot spectra in absorbance or transmittance. Fortunately, many infrared spectroscopy software packages allow you to convert spectra between absorbance and transmittance using Equation 1.7. So, one is not stuck with one set of units or the other. The spectra in this book are plotted in both absorbance and transmittance.

C. The Nature of Molecular Vibrations: Normal Modes

When a molecule absorbs infrared radiation, its chemical bonds vibrate. The bonds can stretch, contract, and bend. This is why infrared spectroscopy is a type of *vibrational spectroscopy*. The vibrational motion excited by infrared absorbance is complex. Fortunately, the complex vibrational motion of a molecule can be broken down into a number of constituent vibrations called *normal modes*. All mechanical systems have normal modes, and will vibrate at some frequency given the right conditions. For example, when a guitar string is plucked, the string vibrates at its normal mode frequency. The normal mode frequency determines the pitch of the note we hear. When a guitar player makes the string longer and shorter using the frets on the instrument, the normal mode frequency of the string is changed, and the pitch we hear gets lower and higher. Molecules, like guitar strings, vibrate at specific frequencies. Different molecules vibrate at different frequencies because their structures are different. This is why molecules can be distinguished using infrared spectroscopy.

It is easy to calculate the total number of normal modes for a molecule, which gives the total number of vibrations that a molecule possesses. The equations are listed in Table 1.1.

Table 1.1 Normal Mode Formulas

Type of Molecule	Normal Mode Formula	# of Modes in a 3 Atom Molecule
Linear	3N-5	4
Nonlinear	3N-6	3

where
    N = the number of atoms in a molecule

A linear 3-atom molecule such as carbon dioxide (O=C=O) has (3x3)-5 = 4 normal modes. The normal modes of CO_2 are a symmetric stretch, an asymmetric stretch, and two bending vibrations. Water is a nonlinear molecule, and has (3x3)-6 = 3 normal modes. The normal modes of water consist of a symmetric stretch, an asymmetric stretch, and one bending vibration. When a molecule absorbs infrared light and begins vibrating, it gives rise to a peak in the infrared spectrum of the molecule at the wavenumber of light absorbed. When we assign a feature in an infrared spectrum, we label it with the functional and the primary type of vibration the molecule underwent when it absorbed light at that wavenumber.

It may be tempting to think that the number of normal modes a molecule has equals the number of its infrared bands. However, not every vibration of a molecule can be excited by infrared radiation (see below). Thus, the number of normal modes acts as a rough guide to how many infrared bands a molecule may have, but cannot provide an exact number.

III. How Molecules Absorb Infrared Radiation

A. The First Necessary Condition for Infrared Absorption

Earlier in this chapter, it was mentioned that one of the constituents of light is the electric vector. The electric vector interacts with matter to produce infrared absorbances. The nature of the electric vector is shown in Figure 1.1. The positive and negative signs on the electric vector represent the polarity of the electric vector. Note that the polarity of the electric vector changes

8   The Basics of Infrared Interpretation

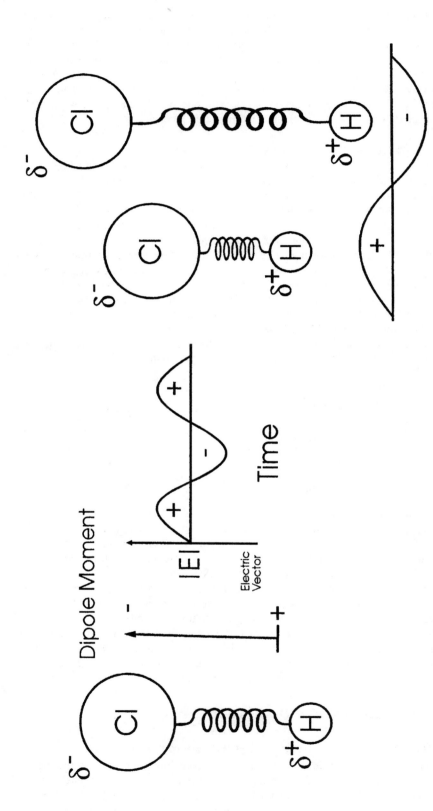

Figure 1.5. A diagram of the interaction of an electric vector with the H-Cl molecule. The δ^+ and δ^- symbols denote the partial positive and negative charges on the hydrogen and chlorine ends of the molecule.

from positive to negative and back again in a cyclical fashion over the course of time. What happens when this alternating polarity electric vector encounters a molecule?

Consider the hydrogen chloride (H-Cl) molecule, whose molecular structure is shown in Figure 1.5. The H-Cl molecule is represented as a "ball and spring" model, where the atoms in the molecule are represented by balls, and the chemical bonds are represented by springs. Because of the *electronegativity* difference between hydrogen and chlorine, the electrons in an H-Cl bond have a much higher probability of being found on the chlorine than on the hydrogen. As a result, the chlorine has a partial negative charge on it as indicated by the δ^- seen in Figure 1.5, and the hydrogen has a partial positive charge on it as indicated by the δ^+. The H-Cl molecule contains two charges separated by a distance. This phenomenon is known as a *dipole moment*. The dipole moment is a measure of the charge asymmetry of a molecule. The magnitude of the dipole moment is calculated from the following equation:

$$\mu = qr \qquad (1.8)$$

where
  μ = magnitude of dipole moment
  q = charge
  r = distance

The size of the positive and negative charges in a dipole are added together to give q. The dipole moment is a vector quantity, having both a magnitude and a direction. Vector quantities can be represented by arrows, where the length of the arrow is proportional to magnitude, and the arrow point gives the direction. A dipole moment vector ($\rightarrow$) is seen in Figure 1.5. If a molecule has more than one chemical bond and the dipoles for each bond are known (the "bond dipole moments"), these are added to give the dipole moment for the entire molecule.

An alternating polarity electric vector is illustrated in Figure 1.5 where **E** is the magnitude of the electric vector. What happens when the electric vector encounters the positive end of an H-Cl molecule? If the electric vector polarity is positive, it will repel the partial positive charge on the hydrogen atom because like charges repel. Consequently, the hydrogen atom will move away from the electric vector and the H-Cl bond will shorten. Once the polarity of the electric vector changes to being negative, it will attract the hydrogen atom because opposite charges attract. Therefore, the H-Cl bond is pulled and it lengthens. The actual change in bond length is only a few percent, and its depiction in Figure 1.5 has been exaggerated for clarity. As the polarity of the electric vector changes over time from positive to negative to positive to negative, the H-Cl bond will get shorter and longer and shorter and longer. During this process, the molecule vibrates at the same frequency as the electric vector, and the energy of the light beam is transferred to the molecule. An *infrared absorbance* takes place. We can draw an analogy between infrared absorbance and a chemical reaction as seen in Figure 1.6.

Before the infrared light beam and molecule interact, the H-Cl molecule is at rest, and the infrared photon has energy equal to hcW. After the interaction, the photon has been absorbed, and its energy deposited into the H-Cl molecule as bond stretching motion. Energy is conserved in this process since all the photon's energy has been transferred to the molecule as vibrational energy. We detect the absorbance of the photon by a decrease in infrared intensity at the wavenumber of the light absorbed, giving an absorption feature in the infrared spectrum of the molecule. In a sense, the dipole moment is the handle on the molecule that the infrared light grasps so it can interact with the molecule.

Figure 1.5 shows how infrared light can interact with a molecule such as H-Cl that has a permanent dipole moment. What about molecules without a permanent dipole moment? Can they interact with infrared radiation? The answer is yes, and the process by which these molecules interact with infrared light is illustrated in Figure 1.7. The bond dipole moments in CO_2 are equal in magnitude but point in opposite directions. Therefore, they cancel and the molecule has no net dipole moment as seen in Figure 1.8.

H—Cl + **Photon** ⟶ **H—Cl**
At Rest Energy = hcW Vibrationally Excited
  Energy = hcW

Figure 1.6. An overview of how an H-Cl molecule becomes vibrationally excited.

Imagine a positive polarity electric vector encountering the negative end of a CO_2 molecule as seen in the left-hand side of Figure 1.7. The electric vector will attract some of the molecule's electrons toward the end of the molecule, making the charge distribution of the molecule asymmetric. Since a dipole moment is a measure of charge asymmetry, the molecule will now have a dipole moment. This dipole moment is called an *induced dipole moment*, because the interaction of the electric vector with the molecule induces the dipole to form. The induced dipole moment is temporary; it exists only when the molecule is interacting with an electric vector.

Once the polarity of the electric vector switches to negative, some of the electrons of the molecule are repelled by it, and they will move toward the opposite end of the molecule. This results in an induced dipole moment opposite in direction than that previously induced. The pushing and shoving of electrons to either end of the CO_2 molecule by the alternating polarity electric vector induces dipoles with which the infrared radiation can interact. The molecule vibrates at the same frequency as the light; energy is transferred from the light beam and is deposited into the molecule as stretching energy. The infrared absorbance process is the same as for H-Cl.

What is the common feature of the vibrations of HCl and CO_2 that allows them to be excited by infrared radiation? Both molecules possess a vibration during which the dipole moment of the molecule changes with respect to distance along the interatomic axis. For example, the dipole moment of H-Cl changes with respect to the H-Cl bond distance during the stretching of this molecule. The first necessary condition for a molecule to absorb infrared light is that the molecule must have a vibration during which the change in dipole moment with respect to distance is non-zero. This condition can be summarized in equation form as follows:

$$\partial\mu/\partial x \neq 0 \qquad (1.9)$$

where:
  $\partial\mu$ = change in dipole moment
  ∂x = change in bond distance

This is a partial differential equation and states that a molecule must possess a vibration that satisfies Equation 1.9 to have an infrared spectrum. Vibrations that satisfy this equation are said to be *infrared active*. The H-Cl stretch of hydrogen chloride and the asymmetric stretch of CO_2 are examples of infrared active vibrations. Infrared active vibrations cause the bands seen in an infrared spectrum.

Molecules can possess vibrations for which the change in dipole moment with respect to bond distance is zero. These vibrations are said to be *infrared inactive*. The symmetric stretch of CO_2 is an example of an infrared inactive vibration. In Figure 1.8, if the oxygen atoms are pulled away from the carbon atom symmetrically, the oxygens are always the same distance from the carbon. At all points during the vibration, the bond dipole moments are of equal magnitude and point in opposite directions, so they always cancel. The value of $\partial\mu/\partial x$ for the entire vibration is zero. The symmetric stretch of CO_2 cannot be excited by infrared radiation and cannot give rise to a band in the infrared spectrum of the molecule.

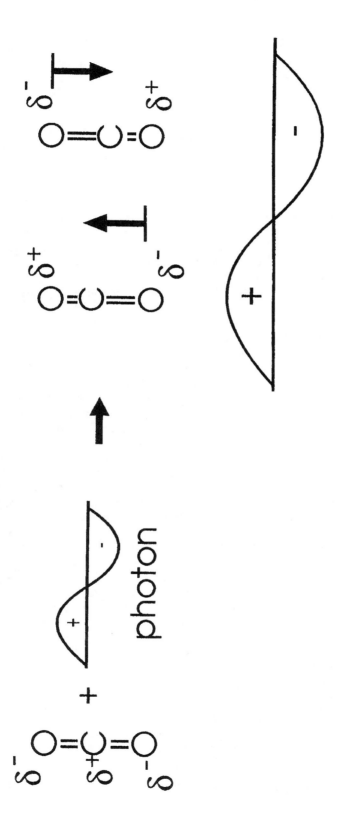

Figure 1.7. An example of an induced dipole moment, the process by which carbon dioxide interacts with infrared radiation.

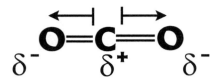

Figure 1.8. The bond dipole moments for the CO_2 molecule. Note how they are equal in magnitude and opposite in direction, canceling each other and giving a net dipole moment of zero.

Equation 1.9 is also important because the size of $(\partial\mu/\partial x)^2$ is directly related to the intensity of infrared absorbance. Infrared bands due to vibrations for which $\partial\mu/\partial x$ is large will be more intense than bands due to vibrations for which $\partial\mu/\partial x$ is small. Homonuclear diatomic molecules, such as N_2 and O_2, have no infrared active vibrations. This is because molecules with two atoms have only one normal mode vibration, a symmetric stretch. The symmetric stretch of a symmetric molecule has a $\partial\mu/\partial x$ of zero. There is no charge asymmetry (a dipole moment) for this vibration, and no way for homonuclear diatomic molecules to absorb infrared light.

B. The Second Necessary Condition for Infrared Absorption

As mentioned above, electromagnetic radiation can be thought of as waves or particles. Recall that particles of electromagnetic radiation are called photons and, according to Equation 1.4, a photon can have any energy. However, the branch of physics known as *quantum mechanics* tells us that molecules can have only certain allowed or quantized vibrational energies. This is a direct result of the atoms in a molecule being in a bound system. Atoms free to move at will can have any energy; atoms bound in a molecule can have only quantized energies.

A diagram of the allowed vibrational energy levels for the water molecule is shown in Figure 1.9. The X-axis is O-H bond distance measured in Angstroms, and the Y-axis is molecular potential energy measured in wavenumber (1 Angstrom equals 1×10^{-10} meter). The curve in Figure 1.9 represents how the potential energy of a water molecule is affected by changes in bond distance. Note that there is a minimum energy at which the molecule is most stable. The bond length at this energy is the *equilibrium bond distance*. To make the bond longer or shorter requires putting energy into the molecule. If the bond is made shorter than the equilibrium bond distance, the molecular energy increases rapidly due to the repulsion of the positively charged nuclei. This is seen in the left-hand part of the potential energy plot. If the bond is made longer than the equilibrium bond distance, molecular energy increases because of the strength of the chemical bond. However, beyond a certain energy the chemical bond will break. That is why the upper right-hand part of the potential energy curve is flat. The thick lines on the potential energy curve represent discrete vibrational energy levels. These energy levels are labeled with what is called a *vibrational quantum number, v*. Typically, the molecule will reside in the ground state ($v=0$) at its equilibrium bond length. The next highest vibrational energy level is $v=1$, and represents the first excited vibrational state of the molecule.

For the O-H stretching vibration of water, the difference between the $v=0$ and $v=1$ levels is about 3500 cm^{-1}. For the molecule to absorb a photon and be promoted to the $v=1$ level, the energy of the photon must equal the energy difference between the ground state and the $v=1$ level. Photons with 3000 cm^{-1} energy are transmitted by water. Photons with 3500 cm^{-1} will be absorbed by water. We have thus deduced the second necessary condition for infrared absorbance to take place: the energy of the light impinging on a molecule must equal a vibrational energy level difference within the molecule.

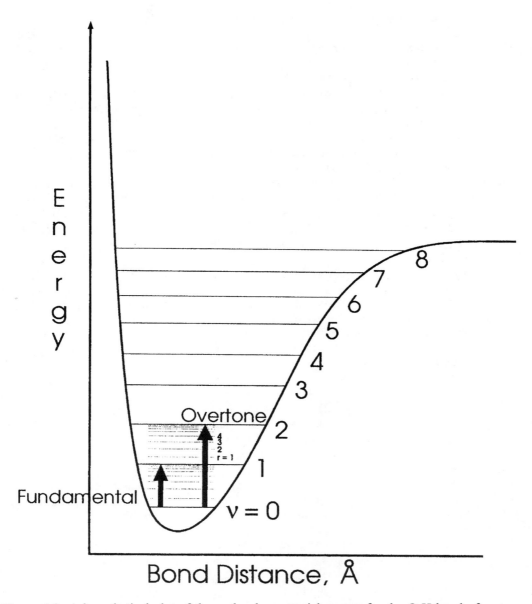

Figure 1.9 A hypothetical plot of the molecular potential energy for the O-H bond of water versus bond distance. The thick lines represent quantized vibrational energy levels. The thin lines represent quantized rotational energy levels.

This idea can be summarized in equation form as follows

$$\Delta E_{vib} = hcW \tag{1.10}$$

where
  ΔE_{vib} = vibrational energy level difference in a molecule
  h = Planck's Constant (6.63 x 10^{-34} Joule-second)
  c = the speed of light (3 x 10^{10} cm/second)
  W = wavenumber in cm^{-1}

Remember that hcW is the energy of a photon. Equation 1.10 puts a restriction on the infrared light that we shine on a sample. If the energy of a photon does not meet the criterion in Equation 1.10, it will be transmitted by the sample. If the photon energy satisfies Equation 1.10, that photon will be absorbed by the molecule (assuming the first condition for infrared absorbance discussed above is met).

Equations 1.9 and 1.10 comprise the two necessary conditions for infrared absorbance to take place. Specifically, a molecule must possess a vibration for which $\partial\mu/\partial x$ is not equal to zero, and the energy of the infrared light we are shining on the molecule must equal a vibrational energy level difference in the molecule. If these two conditions are met, the molecule will absorb infrared light and give rise to a band in the infrared spectrum of the molecule.

If we look closely at Figure 1.9, we can see vibrational energy levels with vibrational quantum numbers of 0, 1, 2 and so on. At room temperature, the vast majority of molecules are in the V=0 or *ground vibrational state*. When a molecule absorbs infrared light and is excited to the first vibrational energy level (v=1), it is said to undergo a *fundamental transition*. A fundamental transition is indicated by the v=0 to v=1 arrow in Figure 1.9. When a molecule is in the v=1 level, it is said to contain one *quantum* of vibrational energy (hence the origin of the term quantum mechanics).

The vast majority of molecules in the universe have vibrations for which the v=1 level is 4000 to 400 cm^{-1} in energy higher than the v=0 level. This means that the vast majority of molecules in the universe have infrared bands between 4000 and 400 cm^{-1}. Most of the intense features in any mid-infrared spectrum can be assigned to fundamental transitions. The wavenumber positions of mid-infrared fundamental bands correlate well with molecular structure, which is why the mid-infrared part of the electromagnetic spectrum is useful, and why you are reading this book.

Figure 1.9 also shows that a molecule can be excited from the v=0 level to the v=2, 3, 4 and higher levels. This type of transition is called an *overtone transition*. If the molecule is excited to the v=2 level, it contains two quanta of vibrational energy; if it is excited to the v=3 level, it contains 3 quanta of vibrational energy. The vibrational energy levels of a molecule are approximately equally spaced. In Figure 1.9, if the V=1 level is at 3750 cm^{-1}, the v=2 level is at about 7500 cm^{-1} (2x3500). This means that the water molecule can absorb light at ~7500 cm^{-1} and undergo an overtone transition from v=0 to v=2. Overtones can be recognized because they are often at about twice the wavenumber of a fundamental band. However, water's overtone band lies well outside the 4000 to 400 cm^{-1} range where most mid-infrared instruments work. Many molecules have overtone bands between 4000 and 13,000 cm^{-1} in what is known as the *near-infrared* (NIR) part of the electromagnetic spectrum. Because of this, overtones are not very common in the mid-infrared spectra of molecules. Overtones give rise to very weak absorbance bands. A v=0 to v=2 transition is typically 10 times weaker than a fundamental band; a v=0 to v=3 transition is typically 100 times weaker than a fundamental band. When overtone bands show up in mid-infrared spectra, they are typically weak and provide little information about the molecular structure of the molecules in a sample. Therefore, in this book, we will simply point out overtone bands in spectra in which they are seen, but we will not use them to identify chemical structures.

As discussed above, the number of atoms in a molecule helps determine the number of vibrations a molecule possesses. Polyatomic molecules can have many infrared active vibrations. It is possible for infrared radiation to excite two or more vibrations at the same time. For example, the asymmetric stretch of water at ~3750 cm^{-1} is infrared active, as is the O-H bending vibration of water at ~1630 cm^{-1}. If light at 5380 cm^{-1} (3750+1630) interacts with a water molecule, the molecule can absorb that light and both the O-H stretching and bending vibrations will be excited at the same time. This would give rise to a band at 5380 cm^{-1} in the near-infrared spectrum of water. This band is called a *combination band* because it arises from a combination of vibrations. Combination bands are also typically much weaker than fundamental bands and do not provide much molecular structural information. We will mention these bands when we encounter them, but they are not very useful.

IV. The Origin of Infrared Peak Positions, Intensities, and Widths

A. Peak Positions

A large part of the rest of this book will be devoted to studying the wavenumbers at which specific functional groups absorb infrared light. Ultimately, we want to derive an equation that will allow us to calculate what wavenumbers of light a given molecule will absorb. Here is how it is done.

1. The Harmonic Oscillator Model of Molecular Vibrations

The easiest way of modeling molecular vibrations is to imagine the atoms in a molecule as balls, and the chemical bonds connecting them as springs. Such a ball-and-spring model for a diatomic molecule is seen in Figure 1.10.

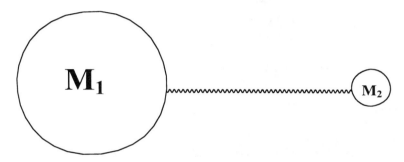

Figure 1.10. A ball-and-spring model of a diatomic molecule. The masses of the two atoms are M_1 and M_2, respectively.

Let us assume for the moment that the masses of the two atoms are M_1 and M_2, respectively, and that the molecule has an infrared active vibration. To make the mathematics simple, let us further assume that when the molecule vibrates, the two atoms move in phase with each other at the same amplitude. These assumptions are part of what is called the *harmonic oscillator* model of molecular vibrations. A real world example of a harmonic oscillator is the swinging pendulum in a grandfather clock. Such motion is simple and "well behaved," and is known as *harmonic motion*.

To describe the molecule's mass in Figure 1.10, we can use a single quantity called the reduced mass. It is calculated as follows:

$$\mu = (M_1 M_2)/(M_1 + M_2) \qquad (1.11)$$

where

μ = reduced mass in Kg (kilograms)
M_1 = mass of atom 1 in Kg
M_2 = mass of atom 2 in Kg

The symbol μ in Equation 1.11 does <u>not</u> represent dipole moment, it represents reduced mass. Unfortunately, the Greek letter μ is used to denote two different quantities, dipole moment and reduced mass, in infrared spectroscopy. Do not be misled by this.

A basic law of physics known as Hooke's Law describes what happens when a spring (or a chemical bond) is stretched. Hooke's Law says

$$F = -kx \qquad (1.12)$$

16 The Basics of Infrared Interpretation

where
F = the restoring force of the spring in Newtons
k = the force constant of the spring in Newtons/cm
x = the displacement of the spring from its equilibrium position in cm

This equation says that the restoring force of a spring depends on two parameters. The first is the force constant (k) of the spring, which is a measure of the spring's stiffness. Second, how far the spring has been pulled (x). In terms of molecules, x is the change in bond length from equilibrium position, and the force constant is a measure of a chemical bond's strength. Equation 1.12 confirms what we already know intuitively about springs, that it takes more force to pull a stiff spring than a weak spring, and it takes more force to pull a spring a long distance than a short distance.

Using Newton's second law

$$F = ma$$

Equation 1.12 can be reformulated as follows:

$$ma = -kx$$

$$-kx = d^2x/dt^2 \, m \tag{1.13}$$

where
F = force in Newtons
m = mass in Kg
a = acceleration meters/sec^2
d^2x/dt^2 = the second derivative of displacement (x) with respect to time (t)

The final equation makes use of the fact that acceleration is the second derivative of displacement with respect to time. This equation is also a partial differential equation. We will solve this equation without proof (consult any elementary text on physical chemistry or quantum mechanics for the mathematical details) to obtain an equation for the wavenumber at which a molecule will absorb infrared radiation.

$$W = \frac{1}{2\pi c} (k/\mu)^{1/2} \tag{1.14}$$

where
W = wavenumber in cm^{-1}
c = the speed of light (3x10^{10} cm/sec)
k = force constant in Newtons/cm
μ = reduced mass in Kg

Equation 1.14 is called the harmonic oscillator equation because it was derived assuming our molecule vibrated with harmonic motion. Recall from Equation 1.3 that

$$\nu = cW$$

Equation 1.14 can then be rewritten as

$$\nu = \frac{1}{2\pi} (k/\mu)^{1/2} \tag{1.15}$$

Equation 1.15 gives the frequency of light that a molecule will absorb, and gives the frequency of vibration of the normal mode excited by that light.

Note that the only two variables in Equation 1.14 are a chemical bond's force constant and reduced mass. These two molecular properties determine the wavenumber at which a molecule will absorb infrared light. Note that a molecule with a large force constant (a stiff or strong chemical bond) will absorb at a high wavenumber, and a molecule with heavy atoms (a large reduced mass) will absorb at a low wavenumber. No two chemical substances in the universe have the same force constants and atomic masses, which is why the infrared spectrum of each chemical substance is unique.

To consider more fully the impact of atomic masses on the positions of infrared bands, consider the two chemical bonds in Table 1.2.

Table 1.2 An Example of a Mass Effect

Bond	C-H Stretch in cm^{-1}
C-^{1}H	~3000
C-^{2}D	~2120

The C-^{1}H bond is a normal carbon-hydrogen bond where the hydrogen has an atomic mass of 1. The C-^{2}D bond represents a carbon-deuterium bond. Deuterium is an *isotope* of hydrogen. It is chemically identical to hydrogen, but has a proton and a neutron in its nucleus, and has an atomic mass of 2. The reduced masses of these two bonds are different, but their force constants are the same. The C-H stretching vibration occurs at ~ 3000 cm^{-1}, whereas a typical C-D bond has a stretching vibration around 2120 cm^{-1}. By simply doubling the mass of the hydrogen atom, the carbon-hydrogen stretching vibration is reduced by over 800 cm^{-1}. A shift in band position caused by a change in reduced mass is called a *mass effect*.

To illustrate how changes in force constants influence infrared band positions, consider the bonds in Table 1.3.

Table 1.3 An Example of an Electronic Effect

Bond	C-H Stretch in cm^{-1}
C-H	~3000
H-C=O	~2750

Again, a typical C-H stretching band is found near 3000 cm^{-1}. When a hydrogen is attached to a carbon with a C=O bond, the C-H stretch band position decreases to ~2750 cm^{-1}. These two C-H bonds have the same reduced mass but different force constants. The oxygen in the second molecule pulls electron density away from the C-H bond, weakening it and reducing the C-H force constant. This causes the C-H stretching vibration to be reduced by ~250 cm^{-1}. Changes in force constant are typically caused by changes in the electronic structure of a molecule, and the shifts in infrared band positions thus caused are called *electronic effects*. When comparing the spectra of similar molecules, such as the spectra of hexane and octane, the slight differences in band positions can be rationalized based on mass and electronic effects.

There are influences external to a molecule such as temperature, pressure, physical state, and chemical interactions that effect the force constant of the molecule and influence the peak position. When comparing spectra of different samples to see if they are the same, all of these variables must be controlled if a legitimate comparison is to be made.

2. Real World Molecules: Anharmonic Vibrations

In our discussion so far, we have assumed that the motions the atoms in a vibrating molecule undergo are harmonic. Although making this assumption made our mathematics easier, it is not a realistic view of the motion of atoms in real vibrating molecules. To help with this picture, imagine a real world analog of the C-H bond, a ping-pong ball connected by a spring to a basketball. This is illustrated in Figure 1.11. If you pull the ping-pong ball and let it go, the ping-pong ball will vibrate at high frequency, bouncing off the more massive basketball. The basketball will move very little, ignoring the motion of the ping-pong ball. This is a good model of a C-H stretch. During this vibration, the light hydrogen atom moves vigorously back and forth while the massive carbon atom hardly moves at all. The carbon and hydrogen do not move with the same amplitude, and do not move in phase with each other. This type of motion does not meet our definition of harmonic motion, and is called *anharmonic motion*. Anharmonic means not harmonic. Anharmonic motion is the type of motion that really takes place in vibrating molecules. Because of *anharmonicity*, the energy levels calculated using Equation 1.14 are only approximate. One would have to add extra terms to Equation 1.14 to take account of anharmonic motion and to calculate the wavenumbers at which a molecule will absorb infrared light more accurately. Additionally, anharmonicity causes the vibrational energy levels depicted in Figure 1.9 to be unevenly spaced, which is why overtone bands are approximately rather than twice the energy of fundamental bands. Polyatomic molecules contain many bonds, each with its own force constant and vibrational energy.

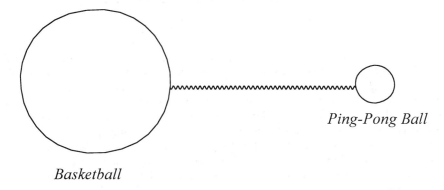

Figure 1.11. A sketch of a basketball connected by a spring to a ping-pong ball. This system is analogous to a C-H bond.

Many of the vibrations in these molecules involve more than one chemical bond. Modeling these vibrations is a challenge since a large number of force constants, reduced masses, and anharmonicities must be known to calculate their vibrational energies. In addition, since all parts of the molecule are connected by "springs," exciting one bond often leads to partial excitation of another bond. This phenomenon is called *vibrational interaction*, and causes vibrational energy levels to be different from those calculated by simple harmonic oscillator theory. An extreme example of vibrational interaction is known as *Fermi resonance*. In Fermi resonance, two vibrations of similar wavenumber and proper symmetry "repel" each other, appearing at wavenumbers above and below where they are normally expected. Vibrations that normally would not appear in the infrared spectrum can "steal intensity" via Fermi resonance and appear in the spectrum. A common example of this is seen in the spectra of aldehydes (Chapter 4). The first overtone of the aldehyde C-H bending vibration coincidentally falls near the energy of the C-H stretching vibration. Normally, the bending overtone would not have enough intensity to appear in the spectrum. However, because of Fermi resonance, the bending overtone steals intensity from the C-H stretching vibration and appears in the spectrum. Additionally, the band positions of these two vibrations are altered, and the intensity of the C-H stretch is less than would be expected. Fermi resonance is rare, but we will see examples of it in this book. Fermi resonance can be used to understand why bands sometimes do not appear at the expected wavenumber or with the expected intensity.

B. The Origin of Peak Intensities

So far, we have thoroughly explored what causes different molecules to have infrared bands at different wavenumbers. However, we have said very little about peak intensities. Peak intensities are useful in infrared spectral interpretation because one can distinguish between the spectra of different functional groups based on peak intensity. In Figure 1.3, which is the infrared spectrum of a pure sample of polystyrene, the different vibrations of the different functional groups in the molecule give rise to bands of differing intensity. This is because $\partial\mu/\partial x$ is different for each of these vibrations. For example, the most intense band in the spectrum of polystyrene is at 698 cm^{-1}, and is due to bending of the benzene ring. One of the weaker bands in the spectrum of polystyrene is at 1601 cm^{-1}, and it is due to stretching of the carbon-carbon bonds in the benzene ring. The change in dipole moment with respect to distance for the ring-bending vibration is greater than that for the C-C stretching vibration, which is why the ring-bending band is the more intense of the two.

An additional factor that determines the peak heights in infrared spectra is the concentration of molecules in the sample. The equation that relates concentration to absorbance is Beer's law, which has the following form:

$$A = \varepsilon l c \tag{1.16}$$

where
   A = absorbance
   ε = absorptivity
   l = pathlength
   c = concentration

The absorbance is measured as a peak height, peak height ratio, peak area, or peak area ratio in the infrared spectrum. The pathlength is typically on the order of microns for solids and liquids, and centimeters to meters for gases in infrared spectroscopy. Concentrations can be measured in moles/liter, ppm, %, pressure, or any of a number of other units depending on the type of sample being analyzed.

The absorptivity (ε) is the proportionality constant between concentration and absorbance, and is dependent on $(\partial\mu/\partial x)^2$. The absorptivity is an absolute measure of infrared absorbance intensity for a specific molecule at a specific wavenumber. The absorptivity does change from molecule to molecule and from wavenumber to wavenumber for a given molecule. For example, the absorptivity of acetone and water are different at 1700 cm^{-1}. The absorptivity for acetone at 1690 cm^{-1} is different than at 1700 cm^{-1}. However, for a given molecule and wavenumber, the absorptivity is a fundamental physical property of the molecule, as invariant as its boiling point or molecular weight. Absorbance is a unitless quantity, so the units of ε are usually given in (concentration x pathlength)$^{-1}$. Thus, absorptivity units cancel the units of the other two variables in Beer's law.

For pure samples, concentration is at its maximum, and the peak intensities are true representations of the values of $\partial\mu/\partial x$ for different vibrations. However, in a mixture, two peaks may have different intensities because there are molecules present in different concentrations. For example, the C=O stretching vibration of acetone at 1715 cm^{-1} is the most intense band in the spectrum of pure acetone. However, if acetone is only 1% of a mixture, the C=O stretch will be small compared to bands from the major components in the sample. Once one learns that C=O vibrations are normally intense, seeing a weak C=O can tip one off that there is a small amount of a carbonyl-containing impurity in a sample.

C. The Origin of Peak Widths

In addition to peak positions and peak intensities, peak widths also provide useful information about a sample. Different functional groups give rise to bands with different peak widths, and this property can be used to distinguish between them. Some functional groups give infrared bands so wide that the widths of the bands by themselves give away the presence of that functional group in a sample. In this book, we will consider the peak widths of liquid and solid samples, since these types of samples are by far the most common samples analyzed in infrared spectroscopy. Anyone interested in a discussion of gas phase bandwidths and the interpretation of gas phase infrared spectra is referred to the author's book on the subject listed in the bibliography.

20 The Basics of Infrared Interpretation

In solid and liquid samples, interactions between nearest neighbor molecules are relatively strong. This is because of the high densities and close packing of molecules found in these phases. As an example, the interaction between neighboring molecules of liquid water is illustrated in Figure 1.12. Any interaction between neighboring molecules is called an *intermolecular interaction*. Because of the electronegativity differences between the oxygen and hydrogen atoms in a water molecule, the electrons in the O-H bond spend most of their time near the oxygen atom. This causes the oxygen atom to have a partial negative charge (δ^-) and the hydrogen atom to have a partial positive charge (δ^+). Since opposite charges attract, the hydrogen atom of one water molecule coordinates with the oxygen atom of another water molecule, forming a weak chemical bond called a *hydrogen bond*. Hydrogen bonds are illustrated by dotted lines in Figure 1.12. Weak chemical interactions such as hydrogen bonds affect the electronic structure and hence the force constants of a molecule. In turn, the force constants of a molecule determine at what wavenumber it will absorb infrared light (see Equation 1.14 above).

From place to place in a sample of a solid or a liquid, the number and strength of nearest neighbor interactions can vary. This is shown in Figure 1.12 where water molecules with one and two nearest neighbors are seen. These water molecules are in different *chemical environments*.

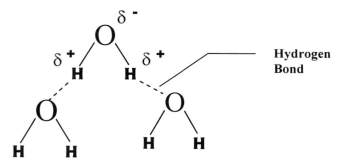

Figure 1.12. The interactions of nearest neighbor molecules in liquid water. The oxygen atoms have a partial negative charge (δ^-) and the hydrogen atoms have a partial positive charge (δ^+). This type of interaction is called hydrogen bonding, as indicated by the dotted lines.

Because the number and strength of hydrogen bonds varies with chemical environment, so do the force constants and the wavenumber at which these molecules absorb infrared light. For example, a water molecule with 1 nearest neighbor may absorb light at 3750 cm^{-1}, while a water molecule with two nearest neighbors may absorb light at 3751 cm^{-1}. Samples with strong intermolecular interactions have more chemical environments than samples with weak intermolecular interactions because electronic effects are greater for strong interactions. Thus, a non-polar molecule like hexane has fewer chemical environments than a polar molecule like water. The more chemical environments a sample has, the more slightly different wavenumbers of light it will absorb. Therefore, broad infrared bands are observed for samples with many chemical environments, and narrow infrared bands are observed for samples with few chemical environments.

In any sample where hydrogen bonding occurs, the number and strength of intermolecular interactions varies greatly within the sample, causing the bands in these samples to be particularly broad. This is illustrated in the spectrum of liquid water, seen in Figure 1.13. Note that the bands of liquid water are hundreds of wavenumbers wide. The spectra of other functional groups that engage in hydrogen bonding, such as alcohols, carboxylic acids, and amines, also show particularly broad bands because of hydrogen bonding. You can make use of these bandwidths to determine if these functional groups are present in a sample.

For molecules whose intermolecular interactions are weak and whose samples contain few chemical environments, infrared bands are narrow. This is illustrated in the spectrum of benzonitrile seen in Figure 1.14. Note that the bands in Figure 1.14 are about <u>tens</u> of wavenumbers wide, an order of magnitude narrower than the bands of liquid water.

Our discussion of peak widths has assumed that the infrared bands in a spectrum are baseline resolved, i.e., that there is no overlap between peaks. This is an unrealistic assumption. It often occurs in infrared spectra that different functional groups absorb at about the same wavenumber. For example, vibrations due to C=O, C=C, and benzene ring functional groups fall between 1700 and 1600 cm^{-1}. If a sample contains some or all of these functional groups, the bands in this region may be broadened due to band overlap. An example of functional group band overlap is seen in Figure 1.15. In this figure, the pure polystyrene has a narrow band at 1492 cm^{-1}. In the polystyrene/polycarbonate mixture, the 1492 cm^{-1} band has been broadened due to overlap with a polycarbonate band that is centered near 1505 cm^{-1}. The widths of these two bands are larger than the distance by which they are separated. Therefore, they overlap forming a complex band with two maxima and a shared baseline. Trying to unravel and interpret these types of features is one of the biggest challenges of infrared spectral interpretation.

To summarize, the width of infrared bands for solid and liquid samples is determined by the number of chemical environments in a sample. In turn, the number of chemical environments is related to the strength of intermolecular interactions. When hydrogen bonding is present (a strong molecular interaction), the number of chemical environments is great and broad infrared bands are observed. When intermolecular interactions are weak, the number of chemical environments is small, and narrow infrared bands are observed.

D. The Origin of Group Wavenumbers

The field of infrared spectroscopy is close to 100 years old. During that time, millions of spectra of many different types of chemical compounds have been obtained. An important observation made by early researchers, and one that still holds true, is that many functional groups absorb infrared radiation at about the same wavenumber, regardless of the structure of the rest of the molecule. For example, C-H stretching vibrations usually appear between 3200 and 2800 cm^{-1}, carbonyl (C=O) stretching vibrations usually appear between 1800 and 1600 cm^{-1}, and aromatic rings have bands below 1000 cm^{-1}. The positions of these infrared bands are determined by reduced masses and force constants. However, the wavenumber ranges where these functional groups absorb are consistent, despite the effects of temperature, pressure, sampling, or changes in molecular structure in other parts of the molecule. This makes these bands diagnostic markers for the presence of a functional group in a sample. These types of infrared bands are called *group wavenumbers* because they tell us about the presence or absence of specific functional groups in a sample. It is the existence of group wavenumbers and their correlation with molecular structure that makes infrared spectroscopy useful. If peak positions varied widely with small changes in the sample, or if there were no correlation between peak position and structure, you would not be reading this book.

A good group wavenumber must have several characteristics. First, it must be intense. It needs to stand out and be easily seen among the many peaks that can appear in an infrared spectrum. Second, a good group wavenumber should appear in a unique wavenumber range. It should be easily identifiable, and not overlap with bands from other functional groups. Last, a good group wavenumber should appear in a narrow wavenumber range, and have a band position that is relatively insensitive to changes in molecular structure and sample condition.

It is useful to break the mid-infrared spectrum into regions. The hydrogen stretching region from 3700 to 2500 cm^{-1} is so called because C-H, O-H, and N-H stretching vibrations show up at these wavenumbers. The triple bond stretching region occurs from 2300 to 2000 cm^{-1}, so called because C≡C and C≡N vibrations appear in this region. From 2000 to 1600 cm^{-1}, the bands of C=C, C=O, and C=N bonds occur giving rise to the double bond stretching region.

22   The Basics of Infrared Interpretation

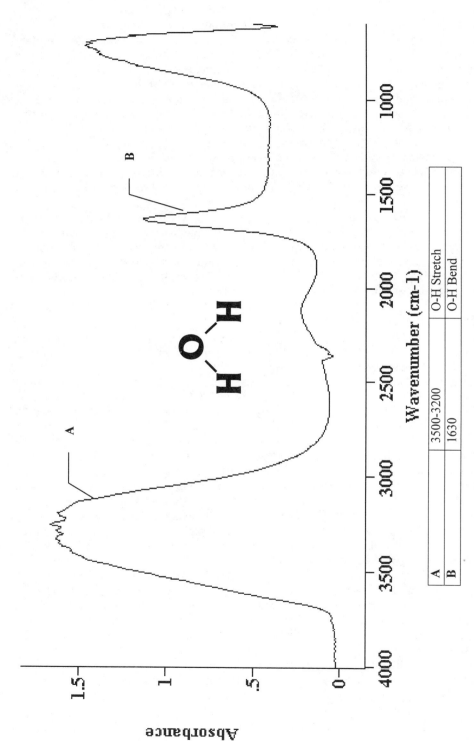

Figure 1.13. The infrared spectrum of liquid water. Notice that the infrared bands are hundreds of wavenumbers wide (attenuated total reflectance sampling technique).

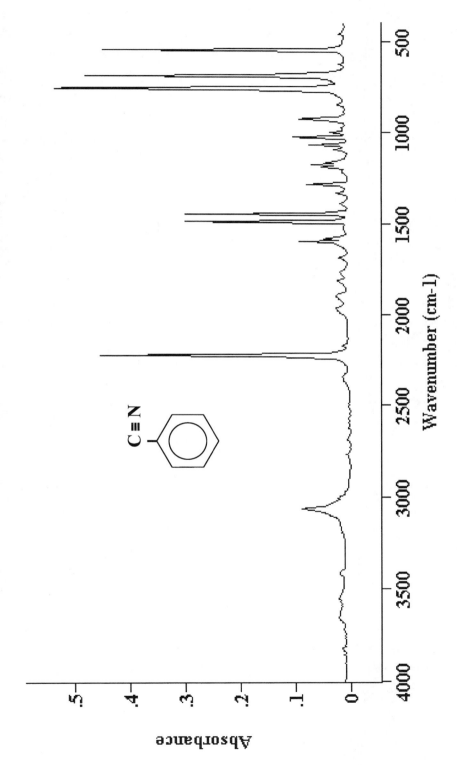

Figure 1.14. The infrared spectrum of benzonitrile. Note that the infrared bands are only tens of wavenumbers wide.

24 The Basics of Infrared Interpretation

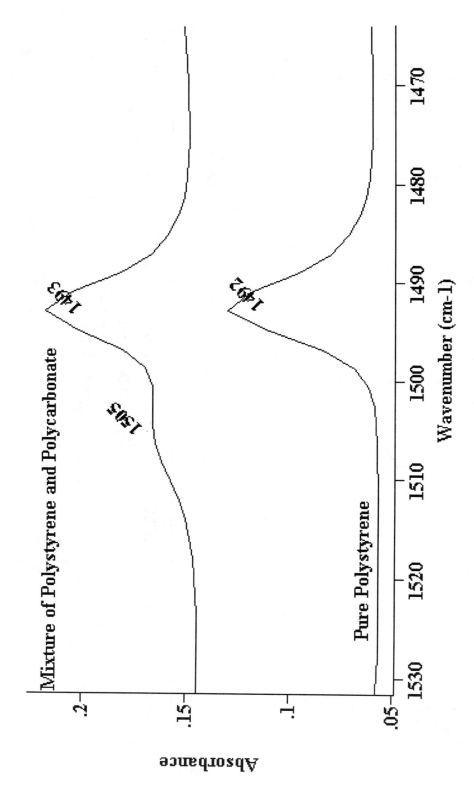

Figure 1.15 Bottom: The infrared spectrum of pure polystyrene. Top: The infrared spectrum of a polystyrene/polycarbonate mixture. Note that the band at 1492 cm^{-1} is broader in the mixture than in the pure sample. This is due to overlap between the polycarbonate band at 1505 cm^{-1} and the polystyrene band at 1492 cm^{-1}.

The region of the infrared spectrum from 1600 to 1000 cm^{-1} is called the *fingerprint region*. This region is notable for the large number of infrared bands that are found there. Many different vibrations, including C-O, C-C, and C-N single bond stretches, C-H bending vibrations, and some bands due to benzene rings are found in this wavenumber region. The fingerprint region is often the most complex and confusing region to interpret, and is usually the last section of a spectrum to be interpreted. However, the utility of the fingerprint region is that the many bands there provide a fingerprint for a molecule. No two molecules in the universe have the same infrared spectrum, and there is no better way to determine whether two samples are the same than by comparing their fingerprint regions. Last, the region from 1000 to 400 cm^{-1} is sometimes called the aromatic region because intense bands due to benzene rings occur there.

V. Infrared Spectral Interpretation: A Systematic Approach

This section will integrate all the information already presented in this chapter into a systematic approach to interpreting infrared spectra.

A. Dealing With Mixtures

Dealing with complex mixtures is one of the problem areas of infrared spectroscopy. The more different kinds of molecules there are in a sample, the more complex the spectrum will be and the harder it becomes to determine what bands are due to what molecules. Ideally, you would only take infrared spectra of pure compounds, since you know all the bands in the spectrum are from one molecule. However, in the real world the vast majority of samples are going to be mixtures. What is to be done about it? The best way of dealing with mixtures is to simplify their composition via purification. Purification methods do not have to be time consuming or expensive. It is sometimes possible to physically separate the components in a mixture. For example, you may be able to pull apart a sample made up of different layers, or observe different color crystals in a sample and tease them apart using tweezers. A filtration, distillation, or recrystallization can clean up solids and liquids. The ultimate way to purify a sample is to chromatographically separate it into its components, collect the components, and then take their spectra individually.

If one cannot simplify the composition of a complex chemical mixture, the next step is to try to simplify its spectrum. This can be done using appropriate software and a technique called *spectral subtraction*. This process can remove the bands of unwanted components from a spectrum. Subtraction involves taking the spectrum of a mixture and subtracting from it the spectrum of a pure compound that is present in the mixture. For example, one could take the spectrum of soapy water, subtract from it the spectrum of pure water, and end up with the spectrum of the soap. This is useful if a spectrum of the pure soap is not available. A more detailed description of spectral subtraction is given in Chapter 9.

Another way to simplify the interpretation of mixture spectra is *library searching*. In this technique, the mixture spectrum is mathematically compared to a collection of known spectra kept in a library. A number called the *hit quality index* (HQI) describes how similar or different the spectra are to each other. A list of the best matches is reported and it is possible to identify a component in a mixture from a library search. In a related process called *subtract and search again*, a library spectrum can be subtracted from the sample spectrum. Then the subtraction result can be searched to try to identify a second component in the sample. Library searching and subtract and search are discussed further in Chapter 9.

Methods of manipulating spectra to overcome overlapped bands found in mixture spectra do exist. A spectral manipulation technique called deconvolution can be used to mathematically enhance the resolution of a spectrum. By causing the features in a spectrum to become narrower, new bands may be seen, allowing unknown components to be identified. Consulting the spectral derivative of the spectrum allows one to determine if new features seen in deconvolved spectra are real or not. Deconvolution is discussed in the author's book on FTIR listed in the bibliography.

If all else fails, the final method of dealing with mixture spectra is the *process of elimination*. The idea here is to use what you know about a sample to tell you what you do not know. First, look for the bands of the atmospheric gases carbon dioxide and water vapor in a spectrum. These tell you nothing about your sample, and should be crossed out and ignored. Next, look for the bands of the molecules you know are in the sample. If there is a solvent, such as liquid water, find its bands and mark them accordingly. Take each component whose presence in the sample you know or suspect, and mark its bands. Finally, you will be left with a collection of bands from unknown molecules. Then, you can use the group wavenumbers discussed throughout this book to determine what functional groups are present in the sample.

B. Properly Performing Identities

In addition to using spectra to identify the structure of unknown molecules, infrared spectra are also used to assess whether two samples are the same. This is called confirming an identity. Identities are used in quality control labs to ascertain whether a raw material is the "right stuff," whether two batches of material are the same, or whether a final product has the right composition to be shipped. This whole process depends on the availability of a spectrum of a reference sample to which the unknown spectrum can be compared. You must be certain of the composition of the reference sample, and its spectrum must be taken under the same conditions as the unknown spectrum. Ideally, these two spectra should be taken on the same instrument, using the same sampling technique, and the same instrumental conditions. Only then can a true comparison be made.

Once the sample and reference spectra are in hand, they must be compared. Plotting the two spectra using the same scale on separate pieces of paper, then overlaying the plots and holding them up to the light is a time honored method of comparing two spectra. This method may seem crude, but it can often times disclose differences between spectra that might not otherwise have been seen. Light boxes can be purchased that shine a light through two overlaid spectrum, and make the comparison process easier. A more modern method of comparing spectra is to overlay them on a computer screen. Finally, programs such as library searches calculate numbers clearly stating how similar two spectra are to each other (see below).

The real question when performing identities is "how close is close enough?" Put more prosaically, how similar do two spectra have to be for the two samples to be considered the same? Recall from above that the data contained in a spectrum are the peak positions, intensities, and widths. The peak positions are determined strictly by the molecules in the sample. If two spectra have all the same peaks, then the two samples contain the same molecules. If the peak positions, intensities, and widths match perfectly, then the two samples have the same molecules in the same concentration. Only then can the two samples be said to have the exact same composition. The degree of spectral similarity that is considered "good" depends upon the application. For example, if you only want to know "is this white powder polyethylene or not?" then a simple comparison of the powder's spectrum to that of polyethylene will give you the answer. If the peak positions match, it is polyethylene. On the other hand, if you are taking spectra of a complex product and are performing identities to determine whether 3 components are present in the right concentrations, you would need your unknown spectrum to exactly match your reference spectrum.

There are ways of quantifying how similar or different two spectra are. You can buy software programs that compare two spectra, and calculate a number that tells how similar the spectra are. Library searching can also do this since the hit quality index (HQI) is a measure of spectral similarity. The challenge is deciding on what HQI is "good enough." You must run a series of good and bad samples, see how variables such as sample preparation, varying operators, and varying instrumental conditions affects the results. All of these things can contribute to two spectra being different even if their concentrations are the same. You should find some number above which all the good samples fall, and below which all the bad samples fall. Once you have established that number, then it can be used to perform "pass/fail" testing on samples.

However, regardless of the results of an automated comparison, you should always <u>visually compare</u> the two spectra in addition to looking at their HQI. Computerized spectral comparisons are not foolproof, and it still takes a human being to check the results of a comparison to ensure that nothing has gone wrong.

C. Infrared Spectral Interpretation: A Systematic 10-Step Approach

When presented with an infrared spectrum, it is easy to become confused by the myriad of peaks present. Many people find the interpretation process frustrating and futile. There is no "magic bullet" that will make IR interpretation easy. However, the author has found that by following a systematic 10-step process one can stay focused and organized, greatly increasing the chances of obtaining accurate information from a spectrum. Here is the 10-step process you can follow when interpreting spectra.

1.) *Use quality data.* If possible, interpret only spectra with a good signal-to-noise ratio, with well resolved peaks, that is free of artifacts. Additionally, the most intense band in a spectrum should have an absorbance less than 2 or a transmittance greater than 100%. Bands more intense than this are totally absorbing and the peaks are truncated. The better the spectrum you start with, the more accurate your analysis will be. If the spectrum you are presented with is not a good one, any number of experimental and software methods can be tried to improve the appearance of a spectrum. Do not settle for lousy data; do everything in your power to ensure your spectrum is a good one before you try to interpret it.

2.) *Avoid mixtures, if possible.* The problems with mixtures and how to handle them were discussed above. Suffice it to say, the more complex the composition of a sample, the more complex its spectrum and the more complex the interpretation job. Anything you can do to simplify the composition of your sample will make your interpretation job easier.

3.) *Use other knowledge of the sample.* Do not interpret your spectrum in a vacuum. Find out as much about your sample as possible beforehand. Ask lots of questions. Have any other chemical analyses been performed on it? What were the results? Where did the sample come from? Who submitted the sample; what do they hope to learn about it? What is the physical state of the sample, its color, and its appearance? Getting answers to all these questions will greatly simplify your interpretation job, and help you narrow down what may be present in a sample.

4.) *Before looking at a spectrum, note its resolution, the sampling method used, and whether any spectral manipulations (subtraction, smoothing, baseline correction, etc.) were performed on the spectrum.* All of these things affect the appearance of a spectrum, and some of them can produce artifacts. If you know this information ahead of time, you can use it while you are interpreting.

5.) *Read the spectrum quickly from left to right noting the presence or absence of the following intense group wavenumbers*:

Band Position in cm^{-1}	**Functional Group**
3500-3200	O-H or N-H
3200-2800	C-H
2250-2000	C≡N, C≡C
1800-1600	C=O
<1000	C=C, Benzene Rings

The idea here is to read the spectrum from left to right like a sentence in a book, and quickly confirm or deny the presence of a large number of important functional groups. This first glance may give you all the information you need to know about a sample.

6.) *Assign the intense bands first* using the tables in this book or other sources of information. The intense bands are intense for a reason; they are important and carry important information; pay attention to them first. We will call the types of infrared bands assigned in steps 5 and 6 *primary bands*.

7.) *Track down the secondary bands of the functional groups whose presence you suspect based on evidence gathered in steps 5 and 6.* The purpose of this step is to confirm the assignments made earlier, and to keep from misassigning bands. For example, it is well known that the C-H stretching vibrations of a benzene ring occur from 3200 to 3000 cm^{-1}. It may be tempting to use this as your only evidence of the presence of a benzene ring in a sample. However, other functional groups absorb in this region (the C-H stretches of alkenes and alkynes), and these bands by themselves do not give the full story. Benzene rings also have bands between 1600 and 1400 cm^{-1}, and from 800 to 600 cm^{-1}. When taken together all these bands are very strong evidence for the presence of a benzene ring in a sample. This is why it is important to track down all the bands due to a functional group, not just the one or two most intense ones.

8.) *Assign other bands as needed.* At this point, you may have successfully assigned some of the bands in a spectrum. However, other bands need attention. There may be secondary bands that need to be found, or primary bands present with which you are not familiar. Use resources such as this book to guide you in the assignment of these "leftover" bands. It is not necessary to assign every wiggle and bump in a spectrum. The most important bands are generally easy to see, and tiny blips in the baseline are generally of little use. Keep this in mind when interpreting spectra.

9.) *Write down the functional groups you think exist in the sample.* As you assign bands, write down the chemical fragments you think are in the sample. Then put together the pieces to come up with proposed chemical structures for the molecules in your sample. You must follow the laws of chemistry when writing proposed structures, remembering that carbon typically has 4 bonds, nitrogen 3, oxygen 2, and hydrogen 1. This step is similar to solving a jig saw puzzle or playing detective in a whodunit. Compare your proposed structure with the spectrum to see if they are consistent with each other. For example, the spectrum may say that you have a benzene ring and an O-H group. If you draw a proposed structure with the two attached to each other, the spectrum can tell you if that assumption is correct or not.

10.) *Get help from spectral atlases, library searching, or interpretation software.* The proper point in the interpretation process to obtain outside help is after you have thoroughly examined the spectrum. It may be tempting to skip the first 9 steps of this process and get help immediately. However, the more you know about your sample, the easier it will be to make sense out of what a library search or interpretation software package may tell you. More information on interpretation aids is given in Chapter 9.

After following these 10 steps, you may still not know the complete chemical composition of a sample. It can be difficult to piece functional groups together into whole molecules using only the information contained in a spectrum. Also, some functional groups do not have good group wavenumbers, and are hard to detect using infrared spectroscopy. This is why using infrared spectroscopy in conjunction with other analysis techniques, such as mass spectroscopy, UV/VIS spectroscopy, NMR, chromatography, and Raman spectroscopy is so important. The more information you have on a sample, the better your conclusions about that sample will be.

VI. Conclusion

The information presented in this chapter represents the <u>science</u> behind infrared spectral interpretation. Much of the rest of the book will represent the <u>art</u> of interpretation. As you will see, every rule stated in this book about group wavenumbers has exceptions, and it is not always obvious when the rules work and when they do not. Consequently, it is very easy to get frustrated when interpreting spectra. With practice, you will get better at interpretation and over time you will learn to quickly spot group wavenumbers. Eventually you may even be able to identify the spectra of molecules at a glance.

Hundreds of infrared band positions are going to be thrown at you throughout the rest of this book. You should by no means try memorizing all of them. The only thing you really need to memorize is the table in step 4 of the interpretation process. You should focus on looking at as many spectra as possible, and use the 10-step process as your guide. Working through the problems at the end of each chapter will also be particularly helpful in learning the process of interpreting spectra. Interpretation is really an exercise in pattern recognition. Once band patterns have established themselves in your brain, you will be on your way to becoming an expert at infrared spectral interpretation.

Bibliography

B.C. Smith, *Fundamentals of Fourier Transform Infrared Spectroscopy*, CRC Press, Boca Raton FL., 1996.

B.C. Smith, *Infrared Gas Analysis: From Instrumentation to Interpretation*, Spectros Associates Press, Shrewsbury MA., 1996.

L.J. Bellamy, *The Infrared Spectra of Complex Molecules*, Wiley, New York, 1954.

R. Silverstein, G. Bassler, T. Morrill, *Spectrometric Identification of Organic Compounds, Fourth Edition*, Wiley, New York, 1981.

D. Lin-Vien, N. Colthup, W. Fately, J. Grasselli, *Infrared and Raman Characteristic Frequencies of Organic Molecules*, Academic Press, Boston, 1991.

N. Colthup, L. Daly, S. Wiberley, *Introduction to Infrared and Raman Spectroscopy*, Academic Press, New York, 1990.

F. Miller, *Lecture Notes for Bowdoin College Infrared Short Course*, Brunswick Maine, 1992.

A. Streitweiser, C. Heathcock, *Introduction to Organic Chemistry*, Macmillan, New York, 1976.

Donald A. McQuarrie, *Quantum Chemistry*, University Science Books, Mill Valley, CA., 1983.

Chapter 2

Hydrocarbons

The term *hydrocarbon* refers to any molecule that contains only hydrogen and carbon atoms. Hydrocarbons are an extremely important family of molecules. Crude oil is a complex mixture of hydrocarbons, and all the materials derived from crude oil, such as gasoline, fuel oil, lubricating oils, and jet fuel are refined mixtures of hydrocarbons. Since the combustion of hydrocarbons supplies the majority of energy consumed by our nation, it is safe to say that this family of molecules is vital to our economy and way life. Therefore, the chemical analysis of hydrocarbons is a large and important field, and the infrared analysis of hydrocarbons provides much important information about these materials.

A *saturated hydrocarbon* is a molecule that contains all the hydrogen atoms it can possibly contain. Saturated hydrocarbons only contain single bonds. Saturated fats contain saturated hydrocarbon chains. *Unsaturated hydrocarbon* molecules do not contain all the hydrogen they can possibly have. They contain double or triple bonds, and these bonds are sites where more hydrogen atoms could be added to the molecule. Unsaturated fats contain unsaturated hydrocarbon chains.

The term *aromatic hydrocarbon* refers to molecules with a special type of carbon-carbon (C-C) bond. These molecules are so named because some of them have pleasant smells. Aromatic molecules are always found in rings. Some of the electrons in each C-C bond in an aromatic ring are delocalized and spread out across all the C-C bonds in the ring. This delocalization gives aromatic molecules a unique stability, chemistry, and unique infrared bands. The most common type of aromatic ring is the benzene ring. Benzene (C_6H_6) consists of 6 carbon atoms bonded together to form a hexagon, with a single hydrogen attached to each carbon atom. Aromatic hydrocarbons are also unsaturated, since they do not contain a full complement of hydrogen atoms.

An *aliphatic hydrocarbon* is any hydrocarbon that does not contain an aromatic ring. There are three major families of aliphatic hydrocarbons, *alkanes*, *alkenes*, and *alkynes*, all of which will be discussed in this chapter.

I. Alkanes

A. Structure and Nomenclature

The simplest type of hydrocarbon is called an alkane. Alkanes consist strictly of carbon-carbon and carbon-hydrogen (C-C and C-H) single bonds. The simplest type of alkane is a linear or "straight chain" alkane. These molecules are sometimes called "normal" or n-alkanes. An example of the chemical structure of a typical straight chain alkane is shown in Figure 2.1.

$CH_3\text{-}CH_2\text{-}CH_2\text{-}CH_2\text{-}CH_2\text{-}CH_3$

Figure 2.1 The chemical structure of hexane (C_6H_{14}), a straight chain alkane.

All straight chain alkanes consist of a string of CH_2 groups, with the string being terminated on both ends by a CH_3 group. A CH_3 group is called a *methyl* group and, when a methyl group ends an alkane chain, it is called a terminal methyl group. The only methyl groups found in straight chain alkanes are terminal methyl groups. The CH_2 structural unit is called a *methylene* group. The only difference between different n-alkanes is the number of methylene groups in the chain. For example, n-hexane has 4 CH_2s, n-octane has six CH_2s, and n-hexadecane has 14 CH_2s. In fact, any n-alkane can simply be described as a CH_2 chain terminated by methyl groups.

Alkanes are said to be saturated molecules because they have as many hydrogen atoms as they can possibly have. Alkanes only contain single bonds; there are no double or triple bonds across which hydrogen can add. Alkanes have such a simple chemical structure that their molecular formulas can be given by the equation C_nH_{2n+2} where N = 0,1,2... . Straight chain alkanes only vary in the number of methylene groups between the terminal methyl groups, and follow the generic structure seen in Figure 2.2.

$CH_3\text{-}(CH_2)_n\text{-}CH_3$

Figure 2.2 The generic molecular structure for straight chain alkanes. N = 0,1,2,... .

For n-hexane, N = 6, and its molecular formula is C_6H_{14}, while for n-decane, N = 10, and its molecular formula is $C_{10}H_{22}$.

B. The Infrared Spectra of Straight Chain Alkanes

1. C-H Stretching Vibrations

The most useful diagnostic bands to determine the presence of methyl or methylene groups in a sample are the C-H stretching vibrations that these groups exhibit. These vibrations typically fall between 2800 and 3000 cm^{-1}. The methyl (CH_3) group gives rise to two C-H stretching bands, the symmetric and asymmetric C-H stretches. A diagram of these vibrations is shown in Figure 2.3.

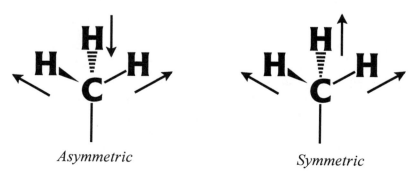

Figure 2.3 The asymmetric and symmetric stretches of a methyl group.

Methyl groups are not planar, and are difficult to represent in two dimensions. In Figure 2.3 the dashed chemical bond represents a C-H bond pointing behind the plane of the page, while the wedge shaped C-H bond is pointing out of the page. The asymmetric C-H stretching motion involves one C-H bond contracting while the other two bonds are lengthening, or two bonds contracting and one lengthening. This band appears at 2962± 10 cm^{-1}. The 20 cm^{-1} range that this vibration falls in is not due to measurement error. This range is the natural variation that the methyl asymmetric stretch exhibits in alkanes due to mass and electronic effects. The spectrum of hexane is shown in Figure 2.5. It exhibits the methyl asymmetric vibration at 2960 cm^{-1}.

The symmetric C-H stretching motion of a methyl group involves all three C-H bonds lengthening or contracting at the same time, as seen in Figure 2.3. The infrared band due to this vibration is found at 2872± 10 cm^{-1}. Hexane displays this band at 2874 cm^{-1}. Note in the spectrum of hexane that the methyl symmetric stretch is less intense and falls at a lower wavenumber than the methyl asymmetric stretch. This is because dipole moment change for the asymmetric stretch is larger than for the symmetric stretch.

The methylene (CH$_2$) structural unit also has symmetric and asymmetric vibrations. Diagrams of these vibrations are seen in Figure 2.4. Fortunately, the methylene group is planar and is easy to represent in two dimensions. The asymmetric C-H stretching vibration for CH$_2$ involves one C-H bond contracting while the other bond is lengthening. This vibration is observed at 2926±10 cm^{-1}, and hexane shows this band at 2926 cm^{-1}. The symmetric methylene stretch involves both the C-H bonds lengthening or contracting at the same time. This band typically appears at 2855± 10 cm^{-1}. Hexane exhibits this band at 2861 cm^{-1}. Note in the spectrum of hexane that the methylene asymmetric stretch is at higher wavenumber and is more intense than the symmetric stretch. This is because the dipole moment change for the asymmetric stretch is larger than for the symmetric stretch.

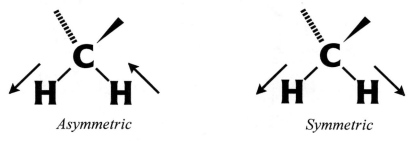

Asymmetric *Symmetric*

Figure 2.4 The asymmetric and symmetric C-H stretching vibrations of the methylene (CH$_2$) structural unit.

As mentioned in Chapter 1, interpreting infrared spectra is an exercise in pattern recognition. In Figure 2.5, there are four closely spaced bands between 2800 and 3000 cm^{-1}. The two bands at higher wavenumber are the most intense. These two intense bands are due to asymmetric stretches; the two less intense bands are due to symmetric stretches. The first and third bands are due to the methyl group; the second and fourth bands are due to the methylene group. This pattern of bands in a spectrum is a strong indication that both methyl <u>and</u> methylene groups are present in a sample. A summary of the C-H stretch group wavenumbers for alkanes is shown in Table 2.1.

Table 2.1. The Wavenumber Ranges for the C-H Stretches of Alkanes (in cm^{-1})

Vibration	Wavenumber Range
CH$_3$ Asymmetric	2962±10
CH$_3$ Symmetric	2872±10
CH$_2$ Asymmetric	2926±10
CH$_2$ Symmetric	2855±10

34 Hydrocarbons

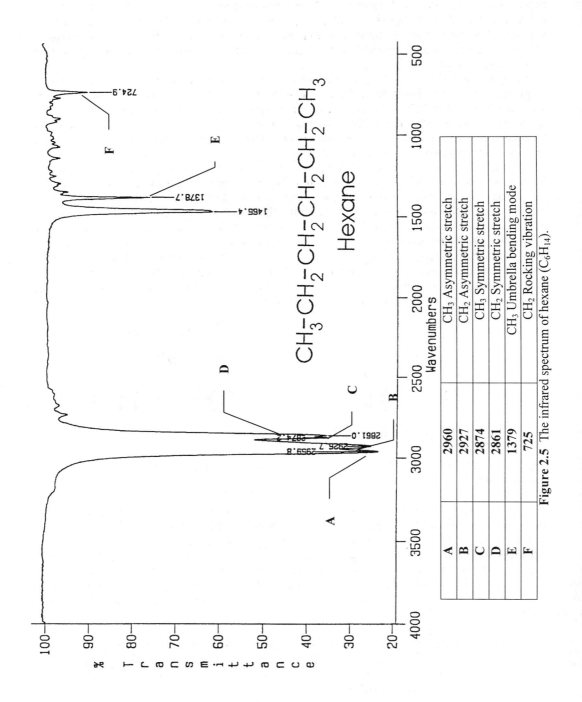

Figure 2.5 The infrared spectrum of hexane (C_6H_{14}).

A	2960	CH_3 Asymmetric stretch
B	2927	CH_2 Asymmetric stretch
C	2874	CH_3 Symmetric stretch
D	2861	CH_2 Symmetric stretch
E	1379	CH_3 Umbrella bending mode
F	725	CH_2 Rocking vibration

A comment on Table 2.1 is in order. These ranges are specific to hydrocarbons. If a methyl or methylene group is attached directly to an oxygen or nitrogen atom, the wavenumber ranges in Table 2.1 may not apply. These nonhydrocarbon C-H stretches will be discussed in later chapters. The presence of nitrogen or oxygen in a molecule can also throw off the expected intensities of the C-H stretches. For example, in the O-CH$_3$ group, the methyl symmetric stretch is more intense than the asymmetric stretch. An unusual intensity pattern in the C-H stretch region, or C-H stretches outside the ranges listed in Table 2.1, should make you suspect there may be nitrogen or oxygen atoms in a molecule. This is why you must be careful about your interpretation of bands in the 2800 to 3000 cm^{-1} region.

Note that the methyl group has two stretching bands, as does the methylene group. If a spectrum exhibits only two C-H stretching bands between 2800 and 3000 cm^{-1}, there is a high probability that it contains only methyl or only methylene groups. Additionally, if only one of the C-H stretches for these groups appears, it will usually be the asymmetric stretch because of its greater intensity. An assignment based just on the asymmetric stretch is often correct. An assignment based on just the symmetric stretch is liable to be wrong, and may indicate the presence of non-hydrocarbon moieties in a molecule.

2. C-H Bending Vibrations

The C-H bonds in straight chain alkanes can undergo bending vibrations. Like C-H stretching vibrations, bending modes are categorized as symmetric and asymmetric. The symmetric and asymmetric bending vibrations of the methyl group are illustrated in Figure 2.6.

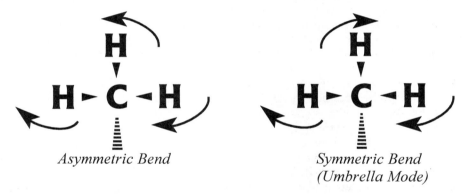

Figure 2.6 The asymmetric and symmetric (umbrella mode) C-H bending vibrations of the methyl (CH$_3$) group.

The easiest way to envision the bending vibrations of a methyl group is to imagine the C-H bonds as the ribs of an umbrella. In Figure 2.6, we are looking down the main axis of the umbrella at the ribs. During the asymmetric bend, one rib of the umbrella is moving in one direction while the other two ribs are moving in the other direction. This vibration appears at 1460±10 cm^{-1}, and appears in the spectrum of hexane (Figure 2.5) at 1465 cm^{-1}. The asymmetric methyl bending band is ambiguous since there is a methylene bending vibration that also falls near 1460 cm^{-1}. Thus, a band in the 1460 cm^{-1} region can indicate the presence of a CH$_3$, a CH$_2$, or both groups. This makes the methyl asymmetric stretch a poor group wavenumber.

The more useful methyl bending vibration is the symmetric stretch. To envision this vibration, imagine the three ribs of the umbrella opening and closing together. This motion is why the methyl symmetric bend is called the "umbrella mode." The term "umbrella mode" will be used throughout the rest of this book to refer to this vibration. Umbrella modes show up at 1375±10 cm^{-1}. Hexane displays this band 1378 cm^{-1}.

The umbrella mode is a very important group wavenumber. The presence or absence of this band is a strong indication of the presence or absence of a methyl group in a sample. As mentioned above, the position of C-H stretching vibrations can be thrown off by the presence of nitrogen and oxygen in a molecule. The position of the umbrella mode is less sensitive to changes in molecular structure than the C-H stretches. Therefore, the umbrella mode can be more reliable than the C-H stretches.

Methylene groups also exhibit two C-H bending vibrations of note. These are illustrated in Figure 2.7.

Figure 2.7 The C-H bending vibrations of the methylene (CH_2) group.

The first of these vibrations involves the H-C-H bond angle getting bigger and smaller, much like the opening and closing of a pair of scissors. This methylene "scissors" vibration appears at 1455 ± 10 cm^{-1}. This vibration is typically overlapped with the methyl asymmetric bend and is not a good group wavenumber. Hexane shows a band at 1465 cm^{-1} that can be assigned as the CH_2 scissors and/or the CH_3 asymmetric bend.

Methylene groups are also capable of "rocking," where the entire group rotates part way in one direction, then part way in the other direction. The axis of rotation for this vibration is the hydrocarbon backbone, and the H-C-H bond angle is maintained during the vibration. The methylene rocking vibration occurs at 720 ± 10 cm^{-1}, and appears in the spectrum of hexane at 725 cm^{-1}. This band usually occurs only in molecules with more than 4 CH_2 groups in a row. Molecules with short hydrocarbon chains do not have a band at 720 cm^{-1}. Thus, the presence or absence of this band can give a crude estimate of hydrocarbon chain length. If the rocking vibration is present in a spectrum, a hydrocarbon chain with at least 4 methylene groups in a row must be present. If this band is absent, the hydrocarbon chains in the sample (if any) must contain 3 or fewer CH_2 groups in a row. However, the CH_2 rocking band does not correlate with the total number of methylenes in a molecule. If a molecule contains many short hydrocarbon chains with less than 4 methylene groups, it will contain many methylene groups, but will not exhibit a CH_2 rocking band. Following in Table 2.2 is a summary of C-H bending vibrations in alkanes.

Table 2.2. The C-H Bending Vibrations of Alkanes (all numbers in cm^{-1})

Vibration	Wavenumber Range
CH_3 Asymmetric bend	1460 ± 10
CH_3 Symmetric bend (umbrella mode)	1375 ± 10
CH_2 Scissors	1455 ± 10
CH_2 Rock	720 ± 10

3. Estimating Hydrocarbon Chain Length from Infrared Spectra

An estimate of hydrocarbon chain length can be made from examining infrared spectra. This is illustrated by the spectrum of hexane in Figure 2.5, and the spectrum of petroleum jelly in Figure 2.8. Petroleum jelly is a mixture of n-alkanes, all of which have more than 20 carbons. Therefore, the alkane chains in petroleum jelly have at least 18 CH_2s in a row. In hexane, the ratio of CH_2 groups to CH_3 groups is 4/2 or 2. In petroleum jelly, the minimum ratio of CH_2s to CH_3s is 18/2 or 9. An examination of the spectra of hexane and petroleum jelly shows that the relative intensity of the CH_3 and CH_2 asymmetric stretches, at ~2962 and ~2926 cm^{-1}, respectively, reflects these ratios.

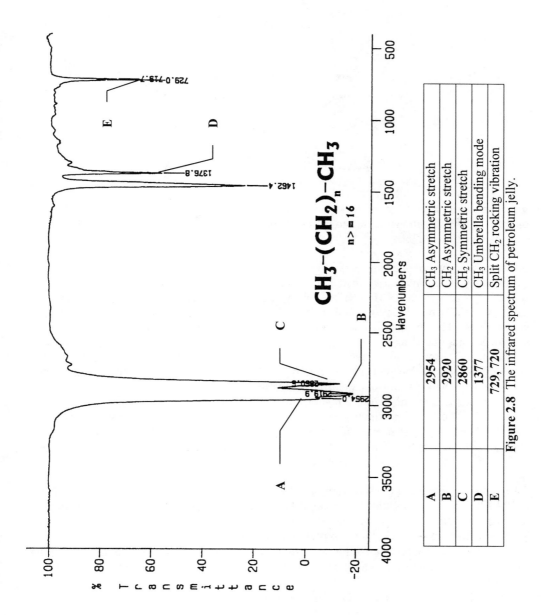

Figure 2.8 The infrared spectrum of petroleum jelly.

A	2954	CH_3 Asymmetric stretch
B	2920	CH_2 Asymmetric stretch
C	2860	CH_2 Symmetric stretch
D	1377	CH_3 Umbrella bending mode
E	729, 720	Split CH_2 rocking vibration

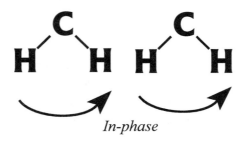
In-phase

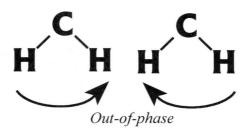
Out-of-phase

Figure 2.9 An illustration of methylene chains rocking in phase (top) and out of phase (bottom). These vibrations cause the split CH_2 rocking band seen at 729 and 720 cm^{-1} in the spectrum of petroleum jelly.

In the spectrum of hexane, the CH_3 and CH_2 asymmetric stretching bands are of approximately equal intensity. In the spectrum of petroleum jelly, the CH_3 band is a small shoulder on a very intense CH_2 asymmetric stretch. These differences in relative intensity are caused by the differences in the number of methylenes compared to the number of methyls in these samples. The absolute length of a hydrocarbon chain can be calculated from an infrared spectrum by first obtaining the spectra of samples of known chain length. A calibration line of the ratio of the absorbances at 2962 and 2926 cm^{-1} versus chain length must then be plotted. Most of the time one does not possess this type of calibration, and only relative hydrocarbon chain length can be inferred from an infrared spectrum.

A closer examination of Figure 2.8 shows other spectral regions that relate to hydrocarbon chain length. The methyl symmetric stretch, which would normally be seen around 2872 cm^{-1}, is not observed in the spectrum of petroleum jelly. This band is obscured by the intense methylene symmetric stretch at 2855 cm^{-1}. Fortunately, there is a clear umbrella mode at 1379 cm^{-1} in the spectrum of petroleum jelly, clearly indicating that there are methyl groups in the sample.

An interesting feature in the spectrum of petroleum jelly is the split CH_2 rocking vibration at 729 and 720 cm^{-1}. This splitting is typical of long chain solid state alkanes, of which petroleum jelly is an example. The splitting arises from closely packed methylene chains. When two adjacent chains begin to rock, they can rock in phase or out of phase with each other as seen in Figure 2.9. Chains that rock out of phase bump into each other, creating a weak interaction that alters the force constant of the vibration just enough to give rise to a second infrared band.

C. The Infrared Spectra of Branched Alkanes

1. Structure and Nomenclature

Not all alkane molecules have straight chains. If an alkane molecule contains a *branch point*, it has more than one C-C chain and is called a *branched alkane*. The structures of the branch points most easily identified from an infrared spectrum are seen in Figure 2.10.

The *gem-dimethyl* group, -C(CH$_3$)$_2$-, involves two methyl groups attached to the same carbon. The gem-dimethyl carbon also has two C-C bonds. Gem-dimethyl groups are internal in a hydrocarbon chain. The other branch points we will discuss end a hydrocarbon chain, and are external to the hydrocarbon chain.

The *isopropyl* group consists of two methyl groups attached to a carbon that also has a C-H bond. The C-H moiety in the isopropyl functional group is called a *methine* group. By definition, methine groups consist of a single C-H bond, and three C-C bonds. All saturated hydrocarbon chains consist of methyl (CH$_3$), methylene (CH$_2$), and methine (C-H) groups. No other functional groups are involved.

The tertiary butyl or *t-butyl* branch point consists of three methyl groups attached to a carbon atom. The term "tertiary" refers to the three C-C bonds, and the term "butyl" means there are four carbon atoms in the *t*-butyl functional group. The central carbon atom in a *t*-butyl group is called a *quaternary carbon* because it has four C-C bonds. The *isobutyl* branch point consists of two methyl groups attached to a methine, to which is also attached a methylene group. Another way to think of an isobutyl group is that it is an isopropyl group bonded to a methylene. As in the *t*-butyl group, the term "butyl" in isobutyl refers to the four carbon atoms in the functional group. Branch points in hydrocarbons give molecular structures more complicated than in straight chain molecules. This is reflected in the complexity of branched alkane spectra compared to unbranched alkane spectra.

Branched and unbranched alkanes can have the same molecular formula. For example, hexane (C$_6$H$_{14}$) has a straight chain structure as seen in Figure 2.5. The compound 2,3-dimethylbutane (also C$_6$H$_{14}$, see Figure 2.11) consists of two attached isopropyl groups, and thus has two branch points. These two molecules have the same chemical formula but different molecular structures, and are called *isomers*.

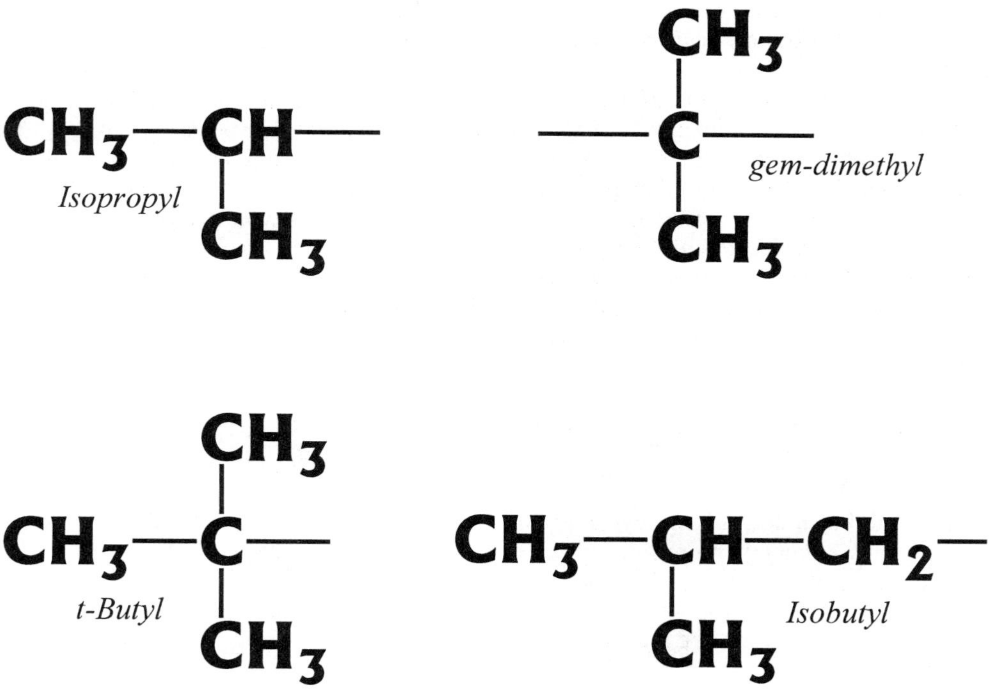

Figure 2.10 Examples of branch points in hydrocarbon chains. Top: isopropyl and *gem*-dimethyl. Bottom: tertiary butyl (*t*-butyl) and isobutyl.

2. Spectra of Branched Chain Alkanes

The simplest branch points that can be observed via infrared spectroscopy are the *gem*-dimethyl and isopropyl groups. Their structure is shown in Figure 2.10. The spectrum of 2,3-dimethylbutane, which consists of two isopropyl groups linked together, is shown in Figure 2.11. The methyl C-H stretching vibrations in Figure 2.11 fall where expected, as does the umbrella mode of the methyl groups. However, close examination of the umbrella mode peak shows it is split, with maxima at 1380 and 1371cm^{-1}. Note that the maxima have about the same intensity. A split umbrella mode with two peaks of about equal intensity is the spectral signature of isopropyl and *gem*-dimethyl groups. The wavenumber range for the two peaks of this split umbrella mode is 1385 to 1365 cm^{-1}. The splitting is caused by vibrational interaction between the umbrella modes of the two methyl groups (think of it as affecting the force constants for the bending vibrations). It is difficult to distinguish between an isopropyl and a *gem*-dimethyl group because both give rise to split umbrella modes with peaks of about equal intensity.

The other two branch points shown in Figure 2.10 are the *t*-butyl and isobutyl group. These groups also give rise to a split umbrella mode, with the two peaks being found between 1395 and 1365 cm^{-1}. However, the intensity ratio of these two peaks is approximately 2:1, with the lower wavenumber peak being more intense. This intensity ratio is illustrated by the spectrum of isooctane seen in Figure 2.12. The umbrella mode peaks are at 1393 and 1366 cm^{-1} and have an approximate intensity ratio of 1:2. The splitting is caused by vibrational interactions between the methyl groups of the branch point. In summary, a split umbrella mode is the spectral signature of a branch point in an alkane, and the relative intensity of the two peaks can help distinguish isopropyl and *gem*-dimethyl branch points from *t*-butyl and isobutyl branch groups.

The C-H stretching region of isooctane bears study. Bands from the asymmetric and symmetric CH$_3$ stretches appear at 2955 and 2870 cm^{-1}. Although isooctane contains a CH$_2$ group, there are no visible methylene stretching modes. They are obscured by the strong CH$_3$ stretching bands. The only other peak in the C-H stretching region is at 2900 cm^{-1}, which corresponds to the stretch of the methine group. This vibration is typically weak, and is often difficult to see beneath stronger methyl and methylene bands. The weak band at 1353 cm^{-1} is due to the C-H bend of the methine group, and is usually not observed. The complexity of this spectrum, and the appearance of bands not found in straight-chain alkanes, flags this spectrum as being of a branched chain alkane. Table 2.3 lists the infrared bands unique to branched chain alkanes.

Table 2.3 C-H Bending and Stretching Bands for Branched Alkanes (all numbers in cm^{-1})

Vibration	Wavenumber
Methine C-H Stretch	~2900 (weak)
Split umbrella mode of isopropyl and *gem*-dimethyl groups	1385 - 1365 (2 bands) Intensity ratio ~1:1
Split umbrella mode of isobutyl and *t*-butyl groups	1395 - 1365 (2 bands) Intensity ratio ~1:2
Methine C-H bend	~1350 (weak)

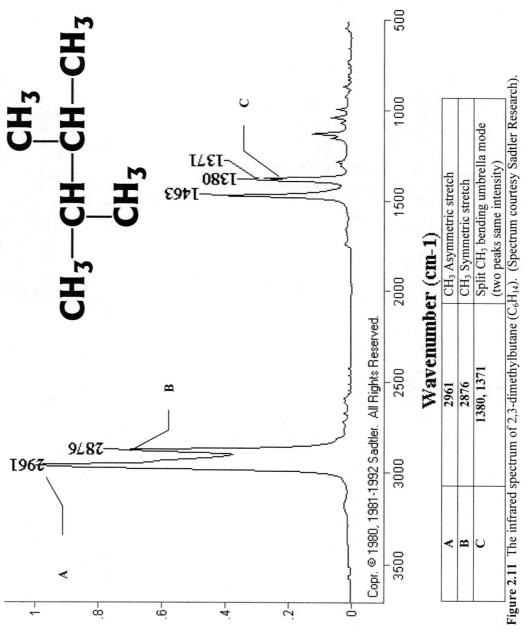

Figure 2.11 The infrared spectrum of 2,3-dimethylbutane (C_6H_{14}). (Spectrum courtesy Sadtler Research).

	Wavenumber (cm-1)	
A	2961	CH_3 Asymmetric stretch
B	2876	CH_3 Symmetric stretch
C	1380, 1371	Split CH_3 bending umbrella mode (two peaks same intensity)

42 Hydrocarbons

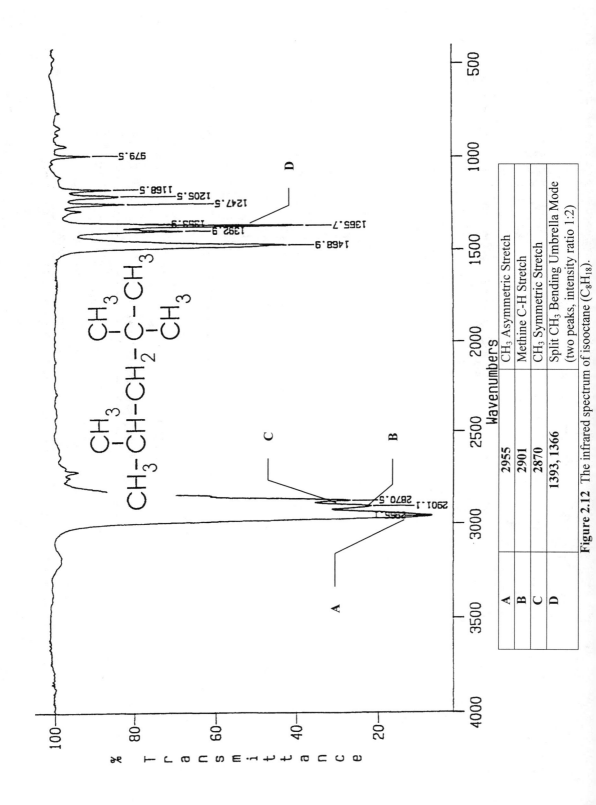

Figure 2.12 The infrared spectrum of isooctane (C_8H_{18}).

A	2955	CH_3 Asymmetric Stretch
B	2901	Methine C-H Stretch
C	2870	CH_3 Symmetric Stretch
D	1393, 1366	Split CH_3 Bending Umbrella Mode (two peaks, intensity ratio 1:2)

II. Alkenes
A. Structure and Nomenclature

An alkene is any molecule that contains a carbon-carbon double bond (C=C). Alkenes are also called *olefins*. The molecular formula for acyclic aliphatic alkenes containing one double bond is C_nH_{2n}. One way to imagine making an alkene is to remove one hydrogen from adjacent methylene groups in an alkane chain. The adjacent carbon atoms will bond to each other, creating a carbon-carbon bond. Since alkenes do not have as much hydrogen as they possibly could have, they are referred to as unsaturated compounds. Unsaturated fats contain hydrocarbon chains with C=C bonds in them. Natural sources of unsaturated fats include vegetable oils such as peanut, sunflower, corn, and olive oils. One application of infrared spectroscopy is measuring the amount of unsaturation in foods by measuring the intensity of alkene infrared bands.

The geometry around a C=C bond is planar and the H-C-H bond angle is about 120°. A double bond has four bonds where hydrogen atoms or other substituents can attach, as seen in Figure 2.13.

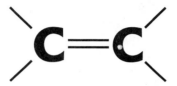

Figure 2.13 The bonding geometry around the C=C bond of an alkene.

There are many ways to arrange substituents around a double bond, leading to many different types of isomers. Figure 2.14 shows some of the possible different substitution patterns. Each "R" in the figure stands for a hydrocarbon chain substituent. The primes (') indicate that each substituent may be the same or may be different.

A *vinyl* group has only one carbon-bearing substituent attached to its double bond, and is found only at the end of a molecule. Hence, molecules containing vinyl groups are sometimes called terminal alkenes.

Vinyl groups are commercially important since many polymers, including polypropylene, polystyrene, polyvinylchloride, polyvinylacetate, and the acrylates are made from vinyl monomers. Billions of pounds of polymers are made each year from vinyl starting materials. One of the important uses of infrared spectroscopy is monitoring the extent of a polymerization reaction by monitoring the intensity of vinyl-group infrared bands.

Two nonhydrogen substituents can be arranged around a double bond in three unique ways. In the *vinylidine* arrangement, both substituents are attached to the same carbon. In the *cis* arrangement the two substituents are on adjacent carbon atoms, but are on the same side of the double bond. In a *trans* ("across") configuration, the substituents are on adjacent carbons, but are on opposite sides of the double bond. Last, tri- and tetrasubstituted double bonds have three and four nonhydrogen substituents, respectively. The infrared spectra of these materials will not only indicate the presence of a double bond, but will allow the substitution pattern to be determined.

B. Infrared Spectra of Alkenes
1. C-H Stretching

Like the C-H bonds of alkanes, the =C-H bonds of alkenes can also stretch and bend, giving rise to useful group wavenumbers. The =C-H stretch of 1-octene (Figure 2.15) is seen at 3077 cm^{-1}. The C-H stretches for almost all unsaturated groups (H-C=C, H-C≡C, and aromatic rings) fall above 3000 cm^{-1}.

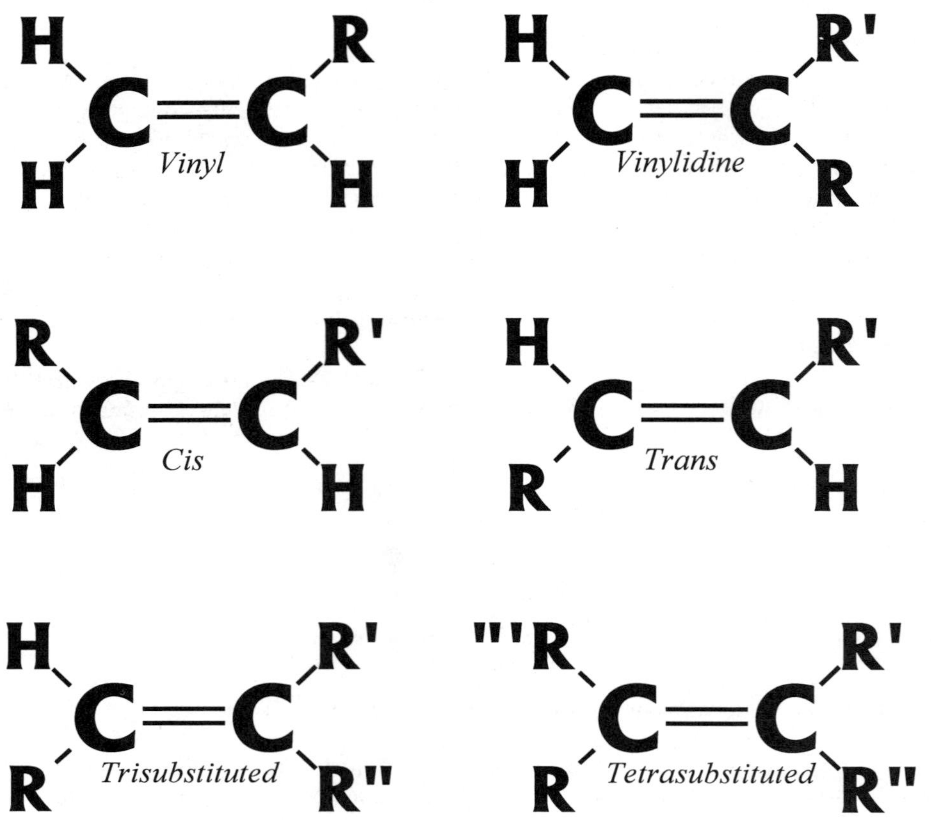

Figure 2.14 The substitution pattern around vinyl, vinylidine, *cis*, *trans*, trisubstituted, and tetrasubstituted double bonds.

The C-H stretches for most saturated groups (CH$_3$, CH$_2$, and CH) fall below 3000 cm^{-1}. If a spectrum contains only C-H stretches below 3000 cm^{-1}, the sample contains totally saturated hydrocarbon chains. If the spectrum contains only C-H stretches above 3000 cm^{-1}, the sample contains totally unsaturated functional groups. If there are C-H stretches above and below 3000 cm^{-1}, the sample contains saturated and unsaturated functional groups. This type of sample is mixed with respect to hydrogen saturation.

The C-H stretches of an alkene can provide more specific information than this. The band at 3077 cm^{-1} in the spectrum of 1-octene is due to the asymmetric C-H stretch of the =CH$_2$ part of the vinyl group. The vinylidine group, since it also contains a =CH$_2$, also exhibits this stretching vibration. The =CH$_2$ asymmetric stretch of vinyl and vinylidine groups generally occurs between 3090 and 3075 cm^{-1}. These groups also show a =CH$_2$ symmetric stretch between 3050 and 3000 cm^{-1}. The other substituted double bond types (*cis*, *trans*, and trisubstituted) also exhibit a C-H stretching vibration between 3050 and 3000 cm^{-1}. The problem with any band in this range is it may be obscured by the C-H stretches of saturated functional groups. Therefore, this latter band is not a good group wavenumber. However, the presence or absence of the =CH$_2$ stretch between 3090 and 3075 cm^{-1} is a good indicator of the presence or absence of vinyl or vinylidine groups in a sample. Table 2.4 summarizes the positions of C-H stretches for saturated and unsaturated functional groups.

Table 2.4. Positions of C-H Stretches for Saturated and Unsaturated Functional Groups
(all numbers in cm^{-1})

Functional Groups	C-H Stretch Position
Saturated (CH$_3$, CH$_2$, CH)	< 3000 cm^{-1}
Unsaturated (C=C, C≡C, Aromatic Rings)	> 3000 cm^{-1}
Vinyl/vinylidine CH$_2$ asymmetric stretch	3090-3075

2. C=C Stretching

One of the more useful group wavenumbers for alkenes is the stretching of the C=C double bond. The general range for this vibration is 1680 to 1630 cm^{-1}. Unfortunately, the intensity of this vibration is variable. Recall from Chapter 1 that the dipole moment change during a vibration determines the absolute intensity of an infrared band. For double bonds that are symmetrically substituted, the dipole moment change is zero, and the C=C stretching band does not appear in their spectra. For double bonds that are nearly symmetrically substituted, the intensity of the C=C stretching band can be weak. On the other hand, if one of the substituents has a large dipole moment, the C=C stretching band can be relatively strong. This means the C=C stretch is useful within limits.

An example of a C=C stretching band is seen in the spectrum of 1-octene in Figure 2.15. The double bond in 1-octene is part of a vinyl group. Its C=C stretch is located at 1641 cm^{-1} and is of medium intensity. In general, the C=C stretch of vinyl, vinylidine, and *cis* alkenes occurs between 1660 and 1635 cm^{-1}. The C=C stretch of *trans*, trisubstituted, and tetrasubstituted compounds occurs between 1680 and 1665 cm^{-1}. Thus, there are two regions in which C=C stretches can be found, and this helps to distinguish between the different substitution patterns. At a minimum, the position of the C=C stretch can distinguish between the *cis* and *trans* isomers of a disubstituted double bond. However, we will need to look at other regions in the spectrum to distinguish between the six types of double bonds whose structures are shown in Figure 2.14.

3. C-H Bending

The most useful group wavenumbers for alkenes are the out-of-plane C-H bends of the hydrogens attached to the double bond. These vibrations appear from 1000 to 650 cm^{-1}. Since a double bond is planar, the C=C and the hydrogens attached to it are in the same plane. The vibration involves the hydrogens bending above and below the plane of the C=C. Out-of-plane =C-H bending vibrations are typically the most intense bands of an alkene, and their position and number will allow us to more completely distinguish between the many different types of double bonds.

The spectrum of 1-octene shows strong bands at 992 and 910 cm^{-1} due to out-of-plane =C-H bending. For vinyl groups in general, these vibrations occur at 990±5 and 910±5 cm^{-1}. Note how sharp and intense these two bands are. Both of these bands must be present for a sample to contain a vinyl group.

The disubstituted double bonds, vinylidine, *cis*, and *trans*, exhibit an out-of-plane C-H bending band at 890±5, 690±50, or 965±5 cm^{-1}, respectively. Trisubstituted double bonds exhibit an out-of-plane C-H bending band at 815±25 cm^{-1}. Tetratsubstituted double bonds have no hydrogens attached to the C=C and do not exhibit C-H stretching or bending bands.

Out-of-plane C-H bending bands, in conjunction with C=C and C-H stretching bands, allow one to distinguish between the many different types of double bond. For example, it is quite easy to distinguish between *cis* and *trans* isomers because *trans* isomers have a C-H bending band at 965 cm^{-1} and *cis* isomers do not. Table 2.5 summarizes the C=C stretching, C-H stretching, and C-H bending bands for substituted double bonds.

46 Hydrocarbons

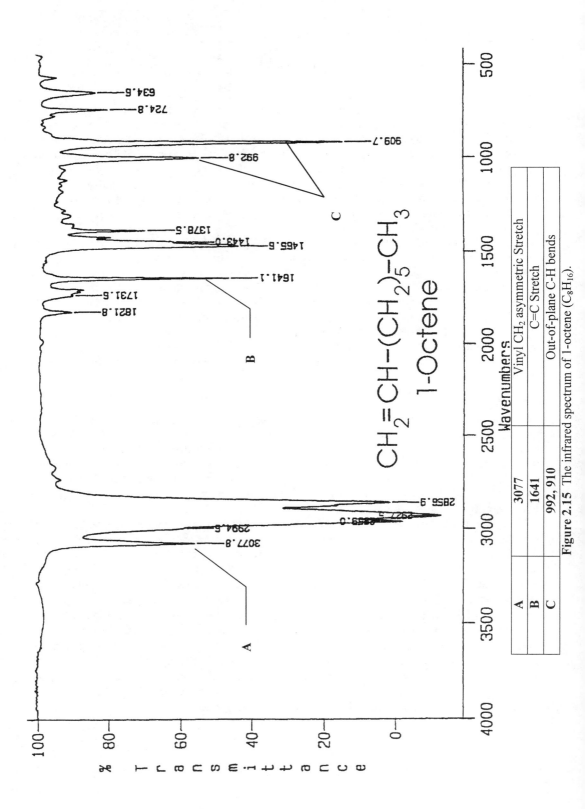

Figure 2.15 The infrared spectrum of 1-octene (C_8H_{16}).

Table 2.5 The C=C Stretching, C-H Stretching, and C-H Bending Bands of Alkenes

Alkene	C-H Stretch	C=C Stretch	C-H Bend
Vinyl	3090-3075	1660-1630	990±5, 910±5
Vinylidine	3090-3075	1660-1630	890±5
Cis	3050-3000	1660-1630	690±50
Trans	3050-3000	1680-1665	965±5
Trisubstituted	3050-3000	1680-1665	815±25
Tetrasubstituted	-	1680-1665	-

III. Alkynes

A. Structure and Nomenclature

An alkyne molecule contains a carbon-carbon triple bond, C≡C. The molecular formula for an aliphatic molecule containing one triple bond is C_nH_{2n-2}. As such, alkynes are another example of an unsaturated functional group. Acetylene (HC≡CH) is the best known example of an alkyne, and alkynes are sometimes named as "substituted acetylenes." There can be either one or two substituents on an alkyne, as seen in Figure 2.16. If R = R' in the disubstituted alkyne, then the molecule is said to be symmetrically substituted.

monosubstituted   *disubstituted*

Figure 2.16 The structure of monosubstituted and disubstituted alkynes.

B. Infrared Spectra of Alkynes

The spectra of monosubstituted alkynes are marked by an intense, high wavenumber C-H stretching vibration between 3350 and 3250 cm^{-1}. This band can be seen in the spectrum of 3,3-dimethyl-1-butyne, shown in Figure 2.17, at 3311 cm^{-1}. This C-H stretching band can be distinguished from O-H and N-H stretching bands that also occur in this region because the alkyne C-H stretch is sharper than O-H or N-H bands. The ≡C-H bond of a monosubstituted alkyne can also bend, and this vibration is called a "wag." The wag band is typically found between 700 and 600 cm^{-1}. The wag of 3,3-dimethyl-1-butyne is seen in Figure 2.17 at 631 cm^{-1}. A disubstituted alkyne does not have any hydrogen attached to the C≡C, and so does not exhibit C-H stretching or bending vibrations.

The last useful group wavenumber for alkynes is the C≡C stretch. This band is found between 2140 and 2100 cm^{-1} for monosubstituted alkynes, and is seen in Figure 2.17 at 2106 cm^{-1}. For disubstituted alkynes, the C≡C stretch is found between 2260 and 2190 cm^{-1}. This band appears in a unique position and, when present, is a strong indication that there is a triple bond in a sample. In addition, the position of this band allows one to distinguish between mono- and disubstituted alkynes. However, if an alkyne is symmetrically substituted, the dipole moment change for the C≡C stretch is zero, and there will be no C≡C stretching band. If an alkyne is almost symmetrically substituted, the C≡C stretch will be weak and may be difficult to see. Therefore, the C-H stretch and wag are more useful group wavenumbers than the C≡C stretch for detection of monosubstituted alkynes. Disubstituted and symmetrically substituted alkynes may be difficult to detect via infrared spectroscopy. Table 2.6 summarizes the useful IR bands for alkynes.

Table 2.6 Summary of Group Wavenumbers for Alkynes (all numbers in cm^{-1})

Substitution	C≡C Stretch	C-H Stretch	C-H Wag
Monosubstituted	2140-2100	3350-3250	700-600
Disubstituted	2260-2190	-	-

48 Hydrocarbons

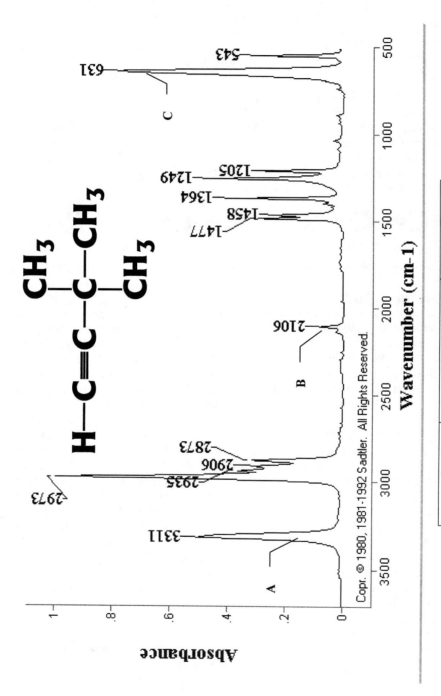

Figure 2.17 The infrared spectrum of 3,3-dimethyl-1-butyne (C_6H_{10}).

IV. Aromatic Hydrocarbons

A. Structure and Nomenclature

The term aromatic hydrocarbon derives from the fact that many of the smaller molecules in this family have strong odors (some more pleasant than others). The prototype of aromatic molecules is the benzene ring, whose two representations are shown in Figure 2.18.

Figure 2.18 The two representations of the benzene (C_6H_6) molecule.

A benzene ring consists of 6 carbon atoms bonded to each other to form a regular hexagon. There is a carbon atom at each vertex of the hexagon, and the C-C-C bond angle is 120°. There is a hydrogen atom attached to each carbon atom, giving benzene the molecular formula C_6H_6. All benzene rings are planar. In the structures seen in Figure 2.18 the carbons and hydrogens are not shown. It is understood that there is a carbon with a hydrogen atom attached to it at each vertex. The circle or lines inside the benzene ring represent some of the electrons involved in carbon-carbon bonding. These electrons are in orbitals that extend above and below the plane of the benzene ring. These electrons circulate around inside the benzene ring and are not part of any specific carbon-carbon bond. These electrons are *delocalized*, the C-C bond order in benzene rings is somewhere between the single bond of an alkane and the double bond of an alkene. For lack of a better term, we can call the carbon-carbon bonds in benzene ring "bonds and a half." This unique bonding gives aromatic rings unique structure, chemistry, and infrared spectra.

There are a variety of molecules that contain benzene rings. One, some, or all the hydrogens on benzene can be replaced by any number of different substituents. There are also different ways of arranging the same substituents on a benzene ring, leading to the existence of many isomeric compounds. An example of a monosubstituted benzene ring is the structure of toluene, shown in Figure 2.19.

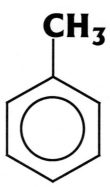

Figure 2.19 The structure of toluene, a monosubstituted benzene ring.

There are no isomers of monosubstituted rings because putting the substituent at any of the six positions on the ring gives six equivalent structures. This can be confirmed by simply rotating the toluene molecule six times through angles of 60°.

There are three unique ways to arrange two substituents on a benzene ring. One must break and make bonds to generate these isomers; they cannot be generated by simply rotating the molecule. The structures of the three isomers of disubstituted benzene rings are shown in Figure 2.20.

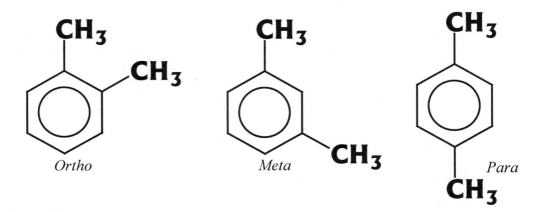

Figure 2.20 The structures of the three different types of disubstituted benzene rings. The structures shown are of the ortho, meta, and para isomers of xylene (dimethyl benzene).

The example molecules pictured in Figure 2.20 are all different isomers of the molecule xylene (dimethyl benzene). In what is known as the *ortho* isomer, the two substituents are on adjacent carbons. The *meta* isomer has the substituents separated by an unsubstituted carbon atom. The *para* isomer shows the substituents at opposite ends of the benzene ring, separated by two unsubstituted carbon atoms.

A discussion of the infrared spectra of all the possible molecules that contain aromatic rings is beyond the scope of this book. We will limit ourselves to mono- and disubstituted benzene rings. However, what we will discover is that an infrared spectrum will not only tell that there is an aromatic ring in a sample, but will also help us determine its substitution pattern.

B. The Infrared Spectra of Aromatic Molecules

1. Overview: The Infrared Spectrum of Benzene

Benzene is the prototype aromatic hydrocarbon. Since it contains 12 atoms, it has 30 (3x12-6) normal modes of vibration. Of these, only a handful are infrared active because of the high symmetry of the molecule. However, substituted benzenes have lower symmetry and can exhibit quite a number of infrared bands. As substituents are arranged around a benzene ring in different ways, the symmetry changes cause some bands to appear, others to disappear, and intensities to vary depending on the dipole moment change for a particular vibration. All of this can make the interpretation of aromatic spectra confusing. The purpose of this section is to give an overview of benzene ring spectra and discuss which of the many benzene ring vibrations are useful for distinguishing between different substitution patterns.

The infrared spectrum of benzene is shown in Figure 2.21. Note how narrow and sharp the bands are in this spectrum. This is a characteristic of aromatic ring spectra, and the sharpness of these bands can be used to distinguish aromatic bands from bands due to other functional groups. Since aromatic rings are unsaturated with respect to hydrogen substitution, the C-H stretches of an aromatic ring fall above 3000 cm^{-1}. Benzene exhibits a trio of bands above 3000 cm^{-1}. In general, the C-H stretches of aromatic rings are found between 3100 and 3000 cm^{-1}. However, a C-H stretch above 3000 cm^{-1} is evidence of unsaturation, but by itself is not enough evidence to prove the existence of an aromatic ring.

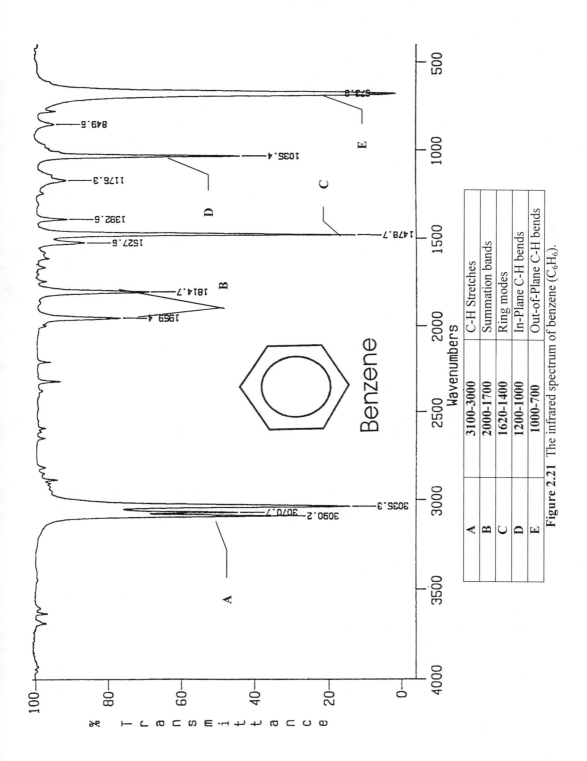

Figure 2.21 The infrared spectrum of benzene (C_6H_6).

A	3100-3000	C-H Stretches
B	2000-1700	Summation bands
C	1620-1400	Ring modes
D	1200-1000	In-Plane C-H bends
E	1000-700	Out-of-Plane C-H bends

Aromatic rings have a series of weak bands in the 2000 to 1700 cm^{-1} region that arise from overtones and combinations of lower wavenumber vibrations. Benzene exhibits these bands at 1959 and 1814 cm^{-1}. In theory, the pattern of these bands can be used to determine the substitution pattern on a benzene ring. The problem with combination bands is that they are often too weak to be of diagnostic use. Typically, a thick sample of pure aromatic hydrocarbon is needed to observe them properly. These bands will not be of use in spectra of mixtures, or if there is not a lot of sample available.

In the 1620 to 1400 cm^{-1} region of aromatic spectra there are bands called *ring modes*. These bands are due to the stretching and contracting of the carbon-carbon bonds in the benzene ring. These are also sometimes called "ring breathing" modes. Benzene has an intense ring mode at 1478 cm^{-1}. The number and intensity of these bands varies with the substitution pattern and the nature of the substituents on a benzene ring. However, ring modes are not ideal group wavenumbers because of the wide range they occupy. Ring modes are useful, however, for distinguishing between alkenes and aromatic rings. If there is a band between 1630 and 1680, it is most likely a C=C (double bond) stretching vibration. If there are bands between 1620 and 1400, they may be benzene ring modes. Ring modes are lower in wavenumber than C=C stretches because the bond order of aromatic carbon-carbon bonds is lower than in alkenes. The concomitant lowering of the force constant causes the difference between the spectra of the two functional groups. Therefore, the 1620 cm^{-1} point in a spectrum is a useful dividing line, denoting the difference between C=C and "C-C bond and a half" stretches.

The region between 1200 and 1000 cm^{-1} in an aromatic spectrum is usually populated by bands due to the C-H bond bending in the plane of the benzene ring. Benzene exhibits one of these bands at 1035 cm^{-1}. The numbers and positions of in-plane C-H bends depend on the benzene ring substitution pattern. However, these bands are typically weak, have wide ranges in which they can appear, and are not useful group wavenumbers.

The best bands for determining the presence of an aromatic ring in a sample are the out-of-plane C-H bending vibrations. These bands usually show up between 1000 and 700 cm^{-1} and involve the C-H bond bending above and below the plane of the benzene ring. There are as many out-of-plane C-H bending vibrations as there are C-H bonds in a benzene ring. However, many of these are weak. Fortunately, the C-H bending vibration involving all the C-H's bending in phase with each other is very intense, often the most intense band in the spectrum of an aromatic hydrocarbon. The position of this band, combined with other information in a spectrum, is the best way of distinguishing between mono- and disubstituted benzene rings

2. The Infrared Spectra of Mono- and Disubstituted Benzene Rings

The infrared spectrum of toluene (methylbenzene) is seen in Figure 2.22. A series of aromatic C-H stretches appear between 3100 and 3000 cm^{-1}, and ring modes show prominently at 1604 and 1495 cm^{-1}. There are more ring modes in the spectrum of toluene than benzene due to toluene's lower symmetry. The ring mode at 1495 cm^{-1} can be distinguished from the methyl asymmetric bend at 1460cm^{-1} because it is sharper and narrower than the C-H bending band. Aromatic in-plane C-H bending bands appear at 1080 and 1029 cm^{-1}. Most important, an intense out-of-plane C-H bending band appears at 728 cm^{-1}. This is the most intense band in the spectrum of toluene. For monosubstituted benzene rings, out-of-plane C-H bending bands fall between 770 and 710 cm^{-1}, and are often the most intense bands in the spectra of these molecules.

A notable feature in Figure 2.22 is the presence of an intense band at 694 cm^{-1}. This band is due to the bending of the C-C bonds in the aromatic ring, and is called the "ring-bending" band. It typically appears at 690±10 cm^{-1}. The appearance of this band is very symmetry dependent. As will be seen below, its presence or absence, in combination with the out-of-plane C-H bending band, will help distinguish between the different types of mono- and disubstituted benzene rings.

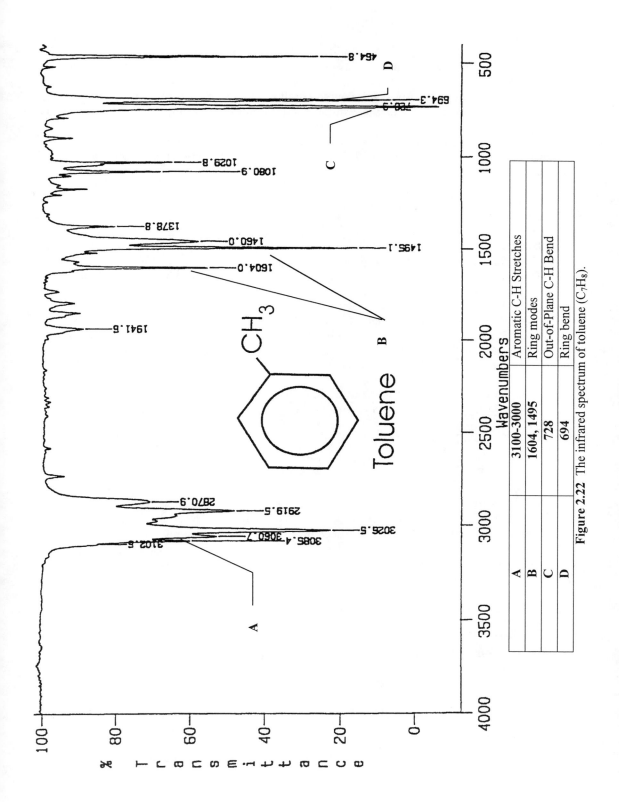

Figure 2.22 The infrared spectrum of toluene (C_7H_8).

A	3100–3000	Aromatic C-H Stretches
B	1604, 1495	Ring modes
C	728	Out-of-Plane C-H Bend
D	694	Ring bend

54 Hydrocarbons

Examples of the spectra of disubstituted benzene rings are seen in Figures 2.23, 2.24, and 2.25. The series of molecules chosen are all different isomers of xylene (dimethyl benzene), and spectra of the ortho, meta, and para isomers are shown, respectively. The spectrum of ortho-xylene is seen in Figure 2.23. This spectrum shows C-H stretches above 3000 cm^{-1} and ring modes at 1604 and 1495 cm^{-1}. The out-of-plane C-H bending band, the most intense in the spectrum, appears at 742 cm^{-1}. Note, however, that there is no ring-bending band at 690 cm^{-1}. Compare this with the spectrum of toluene (Figure 2.22) where the ring bend is the second most intense band. The out-of-plane C-H bending positions for mono- and ortho substituted benzene rings overlap, meaning this band by itself cannot distinguish between these two types of molecules. However, the presence of the ring bend in the monosubstituted spectrum and its complete absence in the ortho-substituted spectrum allows these two types of molecules to be distinguished from each other. The spectral signature of ortho substitution, then, is an out-of-plane C-H bending band between 770 and 710 cm^{-1}, and the absence of an intense band at 690±10 cm^{-1}.

The spectrum of meta-xylene is shown in Figure 2.24. Like other aromatic molecules, its spectrum contains C-H stretches above 3000 cm^{-1} and ring modes at 1612 and 1491 cm^{-1}. The out-of-plane C-H bending band is at 768 cm^{-1}, and is the most intense band in the spectrum. A strong ring-bending band at 691 cm^{-1} is also present in this spectrum, and is typical of meta-substituted benzene rings. Thus, the spectral signature of meta substitution is an out-of-plane C-H bending band between 770 and 735 cm^{-1} and a ring-bending mode at 690±10 cm^{-1}.

The spectrum of para-xylene is shown in Figure 2.25. Like the other xylene isomers, this spectrum has aromatic C-H stretches above 3000 cm^{-1} and an intense ring mode at 1516 cm^{-1}. The most intense band is the out-of-plane C-H bending band at 795 cm^{-1}. There are no features in the 690±10 cm^{-1} region, so there is no ring-bending band present. The spectral signature of para substitution is an out-of-plane C-H bending band between 860 and 795 cm^{-1} with no ring-bending band present. Table 2.7 summarizes the out-of-plane C-H bending and ring bend information for mono- and disubstituted benzene rings.

Table 2.7 Bands for Mono- and Disubstituted Benzene Rings (all numbers in cm^{-1})

Substitution Pattern	Out-of-Plane C-H Bending	Ring Bend (690±10 cm^{-1})
Mono	770-710	Yes
Ortho	810-750	No
Meta	770-735	Yes
Para	860-790	No

The information in Table 2.7 is a powerful tool for distinguishing between mono- and disubstituted benzene rings. However, it is not foolproof. Note that mono- and meta-substituted molecules both have a band at 690 cm^{-1}, and that the range for their out-of-plane C-H bending bands overlap. Thus, if there is a ring-bending band and an out-of-plane C-H bending between 770 and 735 cm^{-1}, it is not obvious if a sample contains a meta-substituted or a mono-substituted benzene ring.

3. Methyl Groups Attached to Benzene Rings

The infrared bands of a methyl group bonded to saturated carbons were discussed earlier in this chapter. However, when a CH$_3$ is directly attached to a benzene ring, the C-H stretching band positions are lowered due to electronic effects. Examples of this can be seen in the spectra of toluene in Figure 2.22 and the xylenes in Figures 2.22 through 2.25. The methyl group C-H stretches in the spectrum of toluene fall at 2919 and 2870 cm^{-1}. These are due to the methyl symmetric stretch and an overtone of the methyl asymmetric bend, which appears via Fermi resonance (see Chapter 1). The methyl asymmetric stretch appears as a weak shoulder on the high wavenumber side of the 2920 cm^{-1} band. For a CH$_3$ directly attached to an aromatic ring, there are two bands at 2925±5 and 2865±5 cm^{-1}, respectively. In a saturated molecule, bands in this range would normally be assigned as methylene C-H stretches.

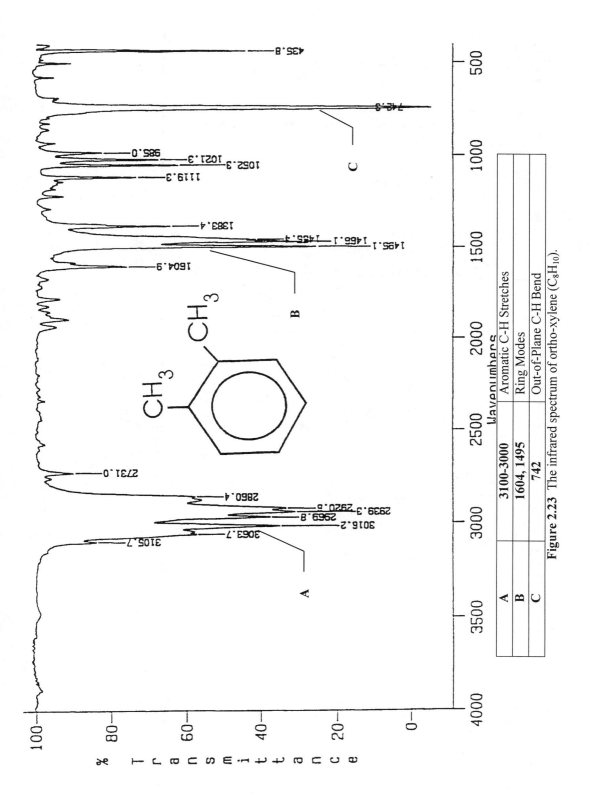

Figure 2.23 The infrared spectrum of ortho-xylene (C_8H_{10}).

	Wavenumbers	
A	3100–3000	Aromatic C–H Stretches
B	1604, 1495	Ring Modes
C	742	Out-of-Plane C–H Bend

56   Hydrocarbons

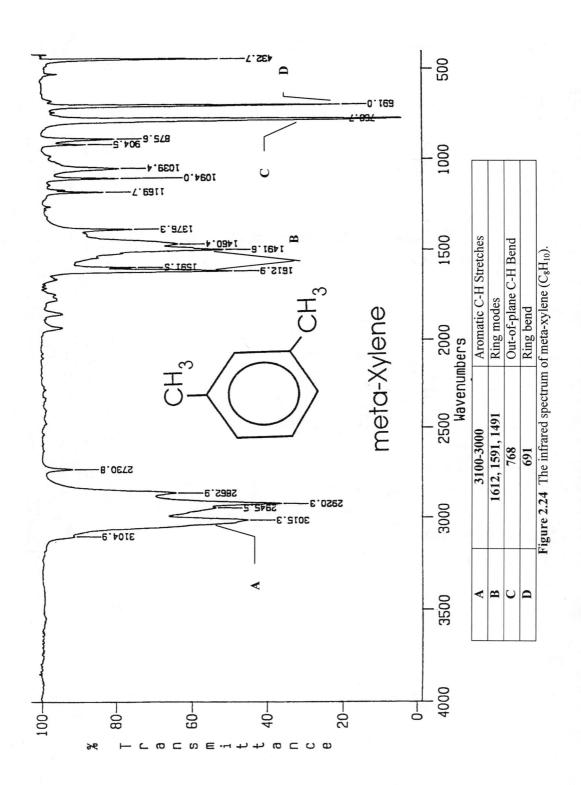

Figure 2.24 The infrared spectrum of meta-xylene (C_8H_{10}).

A	3100–3000	Aromatic C–H Stretches
B	1612, 1591, 1491	Ring modes
C	768	Out-of-plane C–H Bend
D	691	Ring bend

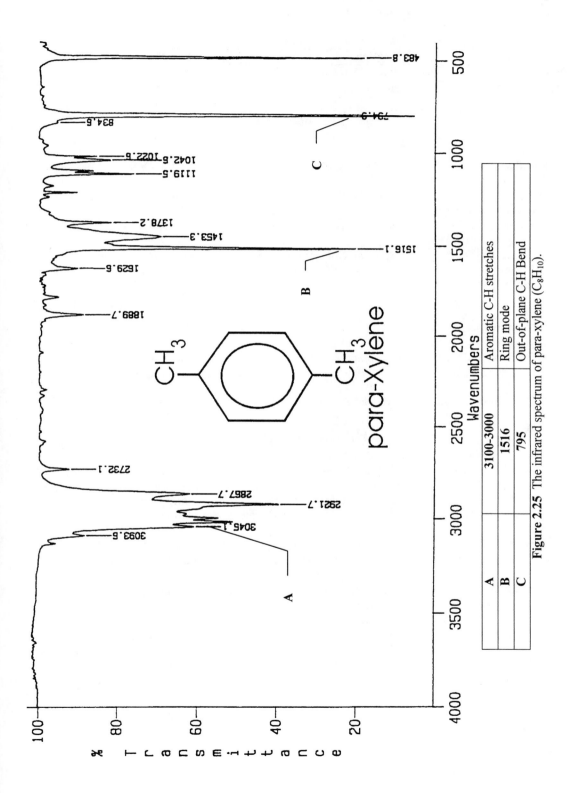

Figure 2.25 The infrared spectrum of para-xylene (C_8H_{10}).

58 Hydrocarbons

However, the C-H stretches above 3000 cm^{-1} indicate there is unsaturation in the sample, which should tip you off that there may be saturated C-H stretches in unexpected positions. Table 2.8 summarizes the group wavenumbers for methyl groups attached to a benzene ring.

Table 2.8 Bands for Methyl Groups Bonded to Benzene Rings (all numbers in cm^{-1})

Vibration	Wavenumber
CH$_3$ Symmetric stretch	2925±5
CH$_3$ Bend overtone	2865±5

Bibliography

L.J. Bellamy, *The Infrared Spectra of Complex Molecules, Third Edition*, Wiley, New York, 1975.
R. Silverstein, G. Bassler, T. Morrill, *Spectrometric Identification of Organic Compounds, Fourth Edition*, Wiley, New York, 1981.
D. Lin-Vien, N. Colthup, W. Fately, J. Grasselli, *Infrared and Raman Characteristic Frequencies of Organic Molecules*, Academic Press, Boston, 1991.
N. Colthup, L. Daly, S. Wiberley, *Introduction to Infrared and Raman Spectroscopy*, Academic Press, New York, 1990.
F. Miller, *Lecture Notes for Bowdoin College Infrared Short Course*, Brunswick, Maine, 1992.
B.C. Smith, *Fundamentals of Fourier Transform Infrared Spectroscopy*, CRC Press, Boca Raton, 1996.
H. Mantsch, D. Chapman, Eds., *Infrared Spectroscopy of Biomolecules*, Wiley, New York, 1996.
A. Streitweiser, C. Heathcock, *Introduction to Organic Chemistry*, Macmillan, New York, 1976.
S. Budavari, Editor, *The Merck Index, 11th Edition*, Merck & Co., Rahway, NJ, 1989.
R.C. Weast, Editor, *The CRC Handbook of Chemistry and Physics*, CRC Press, Boca Raton, 1987.
J. Dean, Editor, *Lange's Handbook of Chemistry*, McGraw-Hill, New York, 1979.

Problem Spectra

The following spectra are presented as an exercise for the reader. Using the techniques and knowledge obtained in this chapter, do your best to predict the complete molecular structure of each sample from its infrared spectrum. The exact positions of prominent peaks are contained in the table accompanying each spectrum. However, these peaks may or may not be useful in the final determination of a structure. The physical state of the sample is also stated (solid or liquid) along with the sampling method used.

All problem set spectra were obtained on modern day FTIR instruments at 4 or 8 cm^{-1} resolution and are of pure materials. It is much easier to learn the basics of interpretation by looking at the simple spectra of pure compounds than to tackle the much more complex problem of analyzing mixture spectra. The assignments learned from interpreting the spectra of simple molecules are useful in understanding the spectra of more complex compounds.

Please realize that the determination of a complete molecular structure in each problem may not be possible. However, by analyzing each of these spectra methodically and patiently, you will be able to learn a lot. The correct structure and a discussion of the interpretation of each problem are included in an appendix at the end of the book. Good Luck!

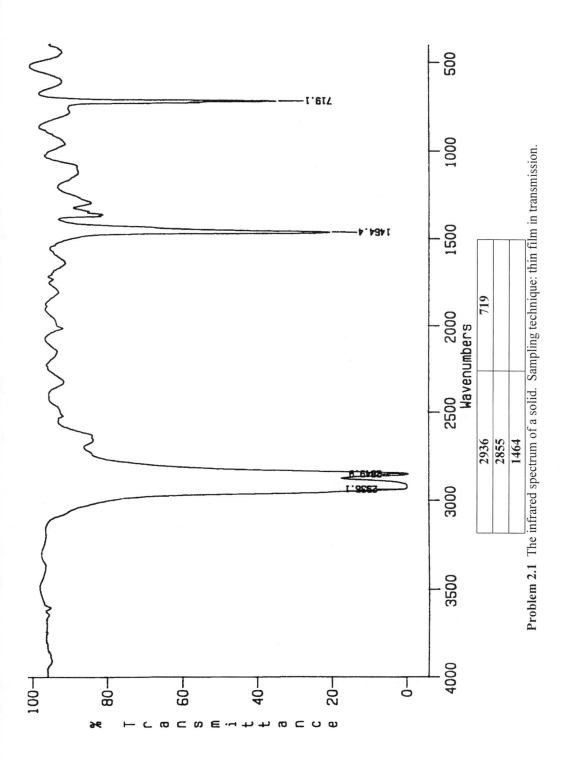

Problem 2.1 The infrared spectrum of a solid. Sampling technique: thin film in transmission.

60 Hydrocarbons

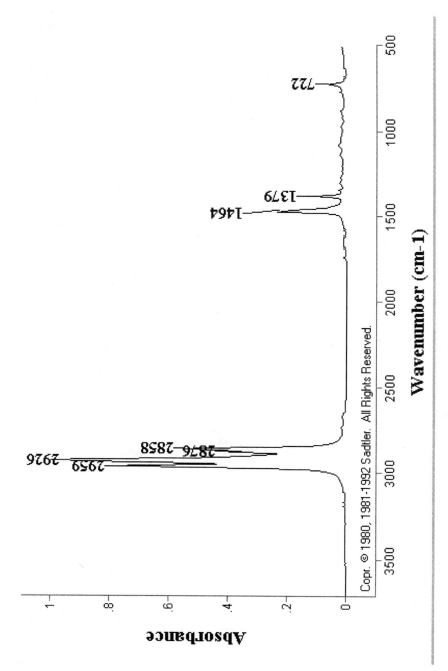

Problem 2.2 The infrared spectrum of a liquid. Sampling method: capillary thin film.

2959	1463
2926	1379
2876	722
2858	

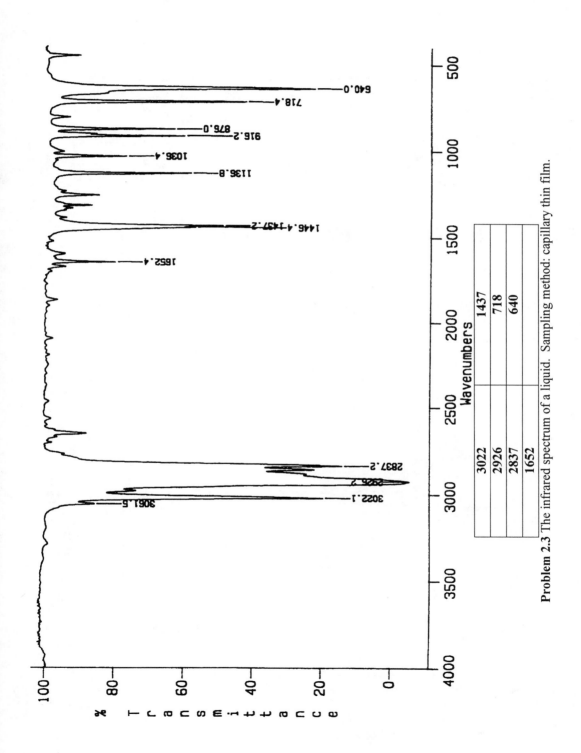

Problem 2.3 The infrared spectrum of a liquid. Sampling method: capillary thin film.

Hydrocarbons

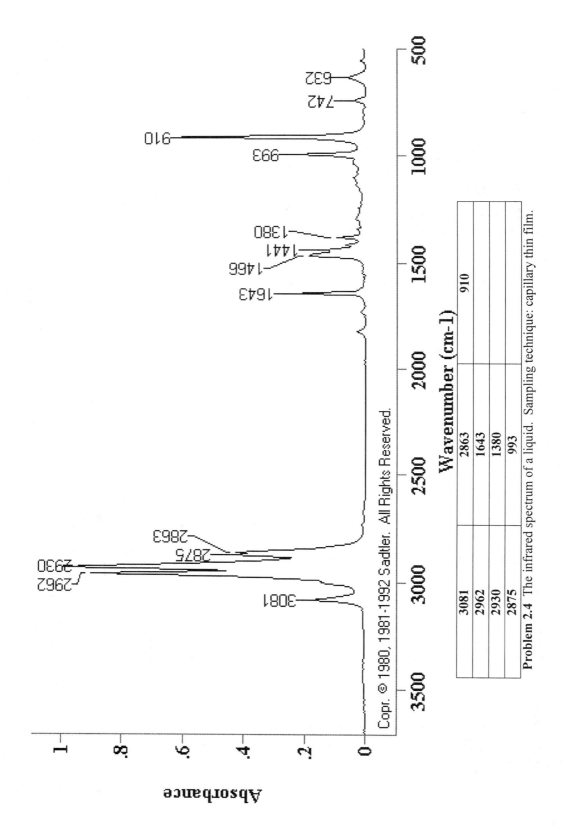

3081	2863	910
2962	1643	
2930	1380	
2875	993	

Problem 2.4 The infrared spectrum of a liquid. Sampling technique: capillary thin film.

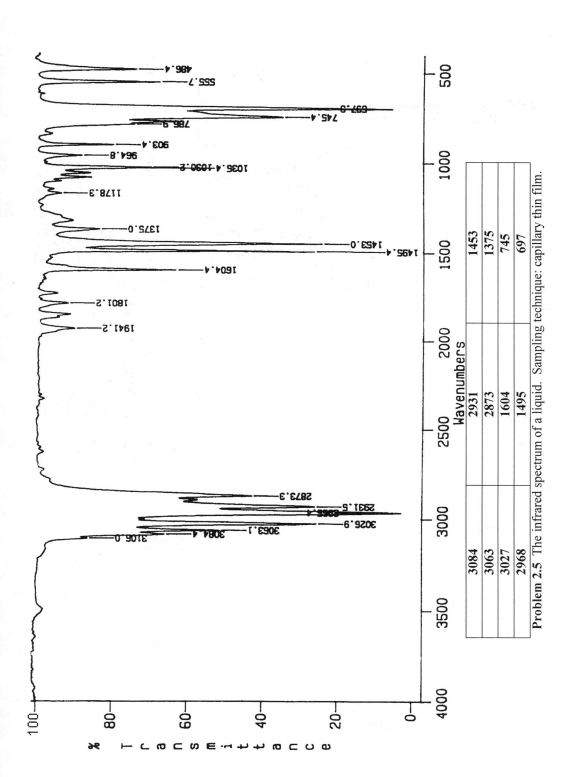

Problem 2.5 The infrared spectrum of a liquid. Sampling technique: capillary thin film.

64　Hydrocarbons

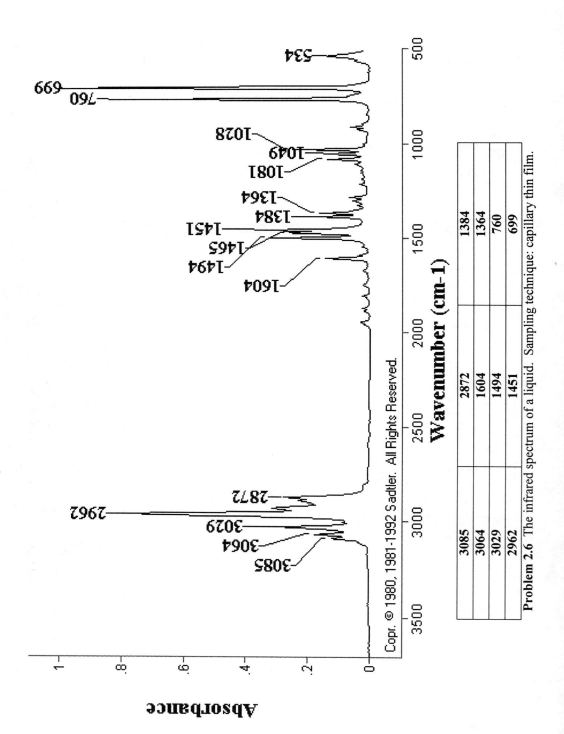

Problem 2.6 The infrared spectrum of a liquid. Sampling technique: capillary thin film.

3085	2872	1384
3064	1604	1364
3029	1494	760
2962	1451	699

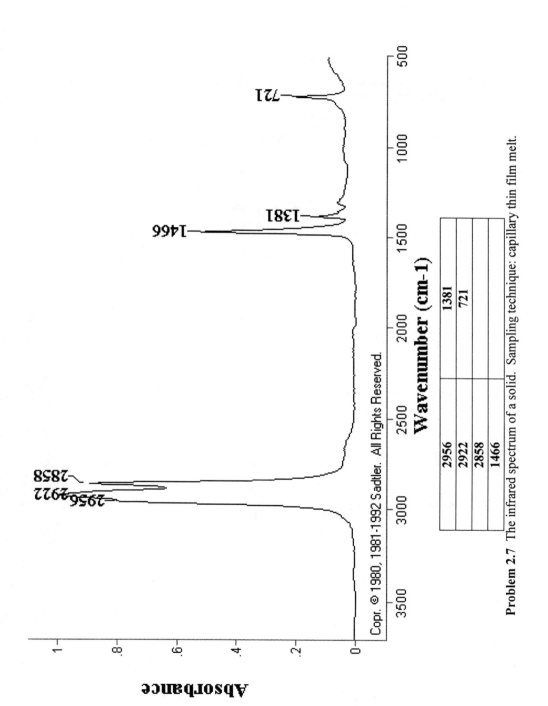

Problem 2.7 The infrared spectrum of a solid. Sampling technique: capillary thin film melt.

2956	
2922	
2858	
1466	
	1381
	721

66 Hydrocarbons

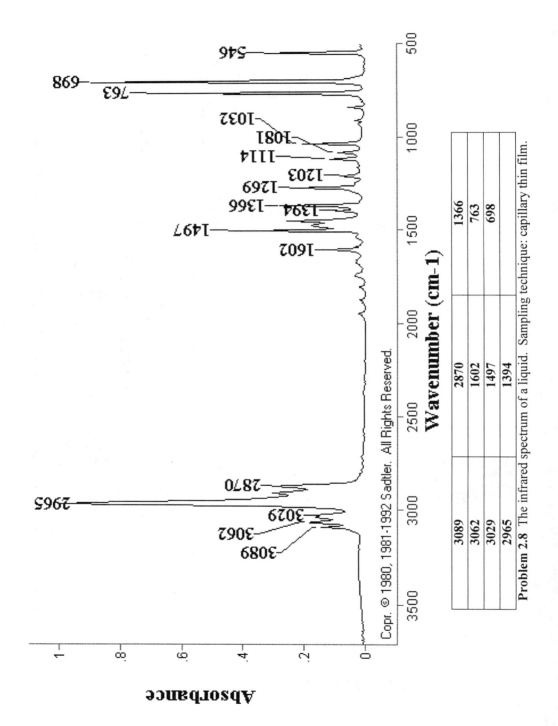

Problem 2.8 The infrared spectrum of a liquid. Sampling technique: capillary thin film.

3089	2870	1366
3062	1602	763
3029	1497	698
2965	1394	

Chapter 3

Functional Groups Containing the C-O Bond

This chapter will discuss the structure, nomenclature, and infrared spectra of chemical functional groups containing the carbon oxygen (C-O) single bond, including alcohols and ethers. All functional groups that contain this bond usually display a C-O single bond stretch between 1300 and 1000 cm^{-1}. This band is usually the most intense in this region, and is sometimes the most intense band in a spectrum. However, the position of the C-O stretch will not always disclose which functional group it is a part of, nor will it always disclose the substitution pattern around the C-O bond. It is often necessary to look at other regions of the spectrum to complete an analysis.

I. Alcohols and Phenols
A. Structure and Nomenclature

An *alcohol* is any molecule that contains a C-OH linkage. The -OH group is called a *hydroxyl* group, and the carbon atom attached to it is called the *hydroxyl carbon*. The structure of a common alcohol, ethanol (also known as ethyl alcohol) is shown in Figure 3.1.

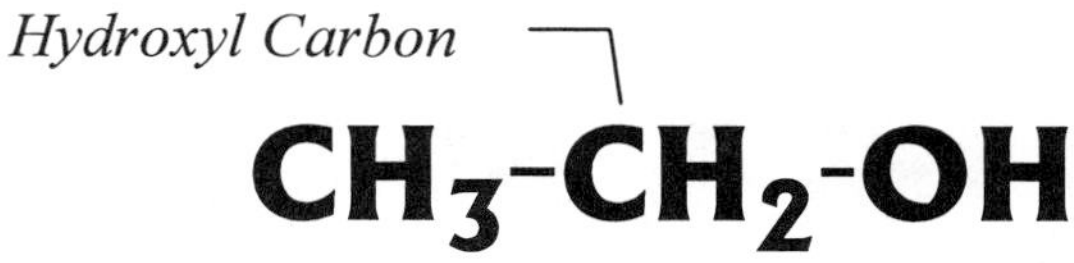

Hydroxyl Carbon

CH$_3$-CH$_2$-OH

Figure 3.1 The structure of ethanol (ethyl alcohol).

A *saturated alcohol* has only saturated carbons attached to the hydroxyl carbon. Ethanol is an example of a saturated alcohol. Saturated alcohols are grouped by the number of saturated carbons attached to the hydroxyl carbon. The structures of the different types of saturated alcohols are shown in Figure 3.2. Ethanol has only one carbon attached to its hydroxyl carbon, so it is called a *primary alcohol*. A *secondary alcohol* has two carbons attached to the hydroxyl carbon, and a *tertiary alcohol* has three carbon atoms attached to the hydroxyl carbon. Chemists often use the symbols 1°, 2°, and 3° to stand for primary, secondary, and tertiary alcohols. Note by definition that a 1° alcohol contains a methylene group, and that a 2° alcohol contains a methine (C-H).

A *phenol* is any molecule with an O-H group directly attached to an aromatic ring. Phenols are examples of *unsaturated alcohols* because aromatic rings are unsaturated with respect to hydrogen substitution. The simplest aromatic alcohol is called phenol, and its structure is shown in Figure 3.3. We will see later in this chapter how the position of the C-O stretching band can help distinguish between saturated and unsaturated alcohols.

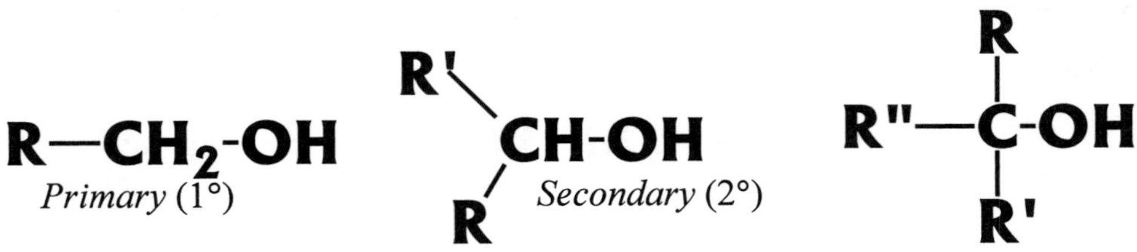

68   The C-O Bond

R—CH₂-OH *Primary (1°)*

R'R\CH-OH *Secondary (2°)*

R"—C(R)(R')-OH *Tertiary (3°)*

Figure 3.2 The structure of primary, secondary, and tertiary alcohols.

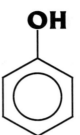

Figure 3.3 The structure of phenol.

B. Infrared Spectra of Alcohols: The Effects of Hydrogen Bonding

The O-H bond of an alcohol is highly polar because of the large electronegativity difference between oxygen and hydrogen. The oxygen atom in an alcohol has a partial negative charge, and the hydrogen atom has a partial positive charge. The positively charged hydrogen on one alcohol molecule can weakly interact with the negatively charged oxygen of a second alcohol molecule. This interaction results in a weak chemical bond called a hydrogen bond. Figure 3.4 shows an example of two alcohol molecules hydrogen bonding.

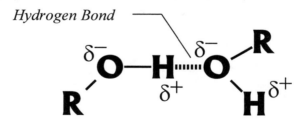

Figure 3.4 An example of two alcohol molecules hydrogen bonding to each other.

In Figure 3.4, the δ^+ and δ^- symbols represent partial positive and negative charges, respectively. The number and strength of hydrogen bonds varies from place to place in a collection of alcohol molecules. This means there are many chemical environments in these samples.

In the discussion of peak widths in Chapter 1, it was shown that the number of chemical environments in a sample contributes to the width of a peak. Because of the large number of chemical environments caused by hydrogen bonding, the infrared bandwidths in alcohols are greater than in other functional groups. Additionally, the infrared bands associated with the O-H group are particularly broad. Figure 3.5 shows the spectrum of an alcohol. The bands at 3342 and 667 cm^{-1} stand out as the widest bands in the spectrum, and must be associated with the O-H group.

The peak width, intensity, and position of alcohol infrared bands are all affected by the extent of hydrogen bonding in a sample. For example, in very dilute solutions where little or no hydrogen bonding takes place, the O-H stretching band of ethanol would appear sharper, weaker, and at higher wavenumber than in Figure 3.5.

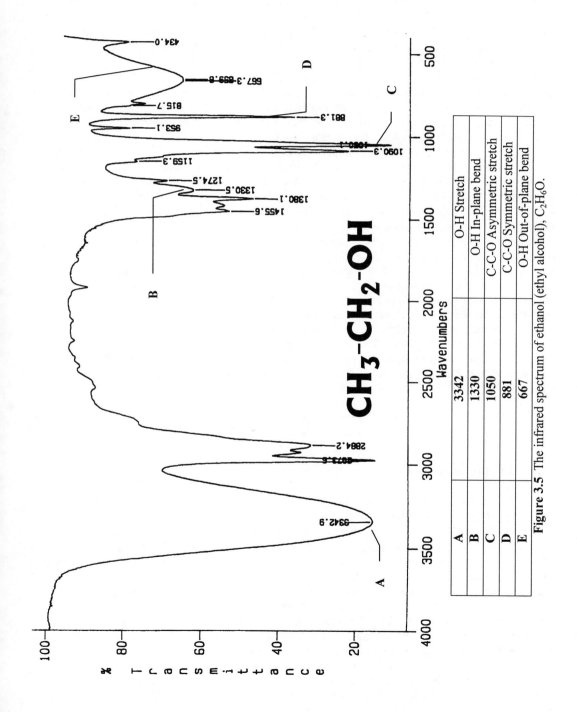

Figure 3.5 The infrared spectrum of ethanol (ethyl alcohol), C_2H_6O.

A	3342	O-H Stretch
B	1330	O-H In-plane bend
C	1050	C-C-O Asymmetric stretch
D	881	C-C-O Symmetric stretch
E	667	O-H Out-of-plane bend

If comparing spectra of alcohols taken under different sampling conditions, any observed spectral differences may be caused by changes in the extent of hydrogen bonding, rather than changes in the composition of a sample.

C. The Infrared Spectra of Saturated Alcohols

As already mentioned, an important spectral feature of alcohols is the width of the bands associated with the O-H group. The most diagnostic of these bands, the O-H stretching vibration, is typically found at 3350±50 cm^{-1}. It is a broad and intense band, often the most intense band in a spectrum. This band occurs in the spectrum of ethanol at 3342 cm^{-1}, as seen in Figure 3.5. This band by itself is a strong indication that there is an alcohol in a sample.

There are two other bands associated with the O-H group, which, in conjunction with the O-H stretch, are a very strong indication that there is an alcohol in a sample. These bands involve the bending of the O-H bond. In-plane bending of the O-H group involves the O-H bond bending in the plane defined by the C-O-H moiety. In Figure 3.1, imagine the O-H group bending in the plane of the page. The in-plane O-H bend occurs at 1350±50 cm^{-1}. Bands due to C-H bending vibrations (such as umbrella modes) also show up near 1350 cm^{-1}. However, any band due to an O-H group will usually be broader than a band due to a C-H group. There are two bands in the spectrum of ethanol located at 1380 and 1330 cm^{-1} that could be assigned as the in-plane O-H bend. However, the band at 1330 cm^{-1} is broader and not as sharp as the band at 1380 cm^{-1}, confirming the assignment of the 1330 cm^{-1} band as the O-H bend (the band at 1380 cm^{-1} is the methyl group umbrella mode).

The second O-H bending vibration is an out-of-plane bend that appears at 650±50 cm^{-1}. This vibration involves the O-H bond bending above and below the plane defined by the C-OH linkage. Imagine the O-H bond in Figure 3.1 bending above and below the plane of the page. The out-of-plane O-H bend can be seen in the spectrum of ethanol at 667 cm^{-1}. Again, other functional groups have bands in this region, but the O-H bend can be distinguished by its width. Sometimes the sharper bands due to other functional groups (aromatic rings, methylenes) can be seen superimposed on the broader O-H bending vibration at ~650cm^{-1}.

The C-O stretch in an alcohol is more appropriately thought of as a C-C-O asymmetric stretch, where the oxygen, hydroxyl carbon, and a carbon adjacent to the hydroxyl carbon are involved in stretching of the C-C-O linkage. A diagram of this vibration is shown in Figure 3.6.

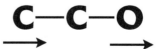

Figure 3.6 The asymmetric C-C-O stretch of an alcohol.

We will use the terms "C-O stretch" and "C-C-O asymmetric stretch" interchangeably throughout the rest of this book.

The position of the C-C-O asymmetric stretch can help distinguish between the three different types of saturated alcohols. For primary alcohols, the C-C-O asymmetric stretch appears from 1075 to 1000 cm^{-1}. This band is seen in the spectrum of ethanol at 1050 cm^{-1}. There is a band in the spectrum of ethanol at 1090 cm^{-1} that might also be assigned as the C-O stretch. However, it is not the most intense band between 1300 and 1000 cm^{-1}, so the assignment of the band at 1050 cm^{-1} as the asymmetric C-C-O stretch is better.

Secondary alcohols exhibit a C-O stretch between 1150 and 1075 cm^{-1}. This is seen in the spectrum of isopropyl alcohol (Figure 3.7) at 1129 cm^{-1}. Isopropyl alcohol is also known as isopropanol or "rubbing" alcohol. There are other bands in its spectrum worth examining; the O-H stretch is at 3349 cm^{-1} and the O-H bending vibrations appear at 1309 and 655 cm^{-1}. Another interesting feature about the spectrum of isopropanol is the split umbrella mode at 1379 and 1369 cm^{-1}. Note that the two halves of the umbrella mode are the same intensity, as expected for isopropyl groups.

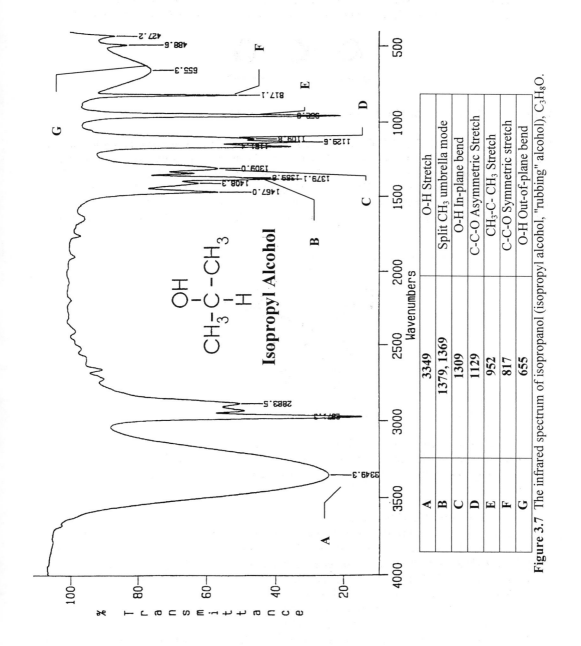

Figure 3.7 The infrared spectrum of isopropanol (isopropyl alcohol, "rubbing" alcohol), C_3H_8O.

A	3349	O-H Stretch
B	1379, 1369	Split CH_3 umbrella mode
C	1309	O-H In-plane bend
D	1129	C-C-O Asymmetric Stretch
E	952	CH_3-C- CH_3 Stretch
F	817	C-C-O Symmetric stretch
G	655	O-H Out-of-plane bend

72 The C-O Bond

Tertiary alcohols give rise to a C-C-O asymmetric stretch between 1210 and 1100 cm^{-1}. Note that this range overlaps that for secondary alcohols (1150-1075 cm^{-1}). Unfortunately, if a C-O stretch falls between 1150 and 1100 cm^{-1}, it is ambiguous whether a sample contains a 2° and 3° alcohol. However, distinguishing between 1° and 2° alcohols based on the position of the C-O stretch is possible.

There are a number of other vibrations alcohols can undergo involving the C-O and adjacent C-C bonds. These give rise to a series of bands between 1000 and 800 cm^{-1}. These vibrations are not useful for distinguishing between different types of alcohols, but they can show up with significant intensity. One of these bands is the symmetric C-C-O stretch, seen in Figure 3.8.

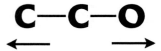

Figure 3.8 The symmetric C-C-O stretching vibration of an alcohol.

For 1° and 2° alcohols, the C-C-O symmetric stretch shows up between 900 and 800 cm^{-1}. In the spectrum of ethanol, it appears at 881 cm^{-1}, and in the spectrum of isopropyl alcohol, it appears at 817 cm^{-1}. Many secondary alcohols are like isopropyl alcohol and exhibit this band near 820 cm^{-1}. Tertiary alcohols have a C-C-O symmetric stretch band at ~1000 cm^{-1}. Bands involving the asymmetric stretch of the C-C-C bonds in an alcohol, where the middle carbon is the hydroxyl carbon, also appear in the spectra of alcohols. This vibration is illustrated in Figure 3.9.

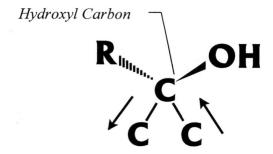

Figure 3.9 The C-C-C stretch of an alcohol. The middle carbon is the hydroxyl carbon.

Many alcohols have an intense band between 1000 and 900 cm^{-1} due to this C-C-C stretch. This band is seen in the spectrum of isopropyl alcohol at 952 cm^{-1}. Since this vibration involves three carbon atoms, short chain alcohols such as ethanol and methanol will not have this vibration.

D. The Infrared Spectra of Phenols

The term phenols is used to describe any molecule where an O-H group is directly attached to a benzene ring. Phenols are a type of alcohol, but are unsaturated alcohols because the carbons in a benzene ring are unsaturated. The chemistry and properties of phenols are different than those of saturated alcohols, and so are their infrared spectra. The infrared spectrum of the molecule phenol (hydroxybenzene) is shown in Figure 3.10.

Like any other O-H containing molecule, the bands due to the O-H group in the spectrum of phenol can be easily picked out due to their width. The band positions of the stretching and bending vibrations of the O-H group are not sensitive to the presence of an aromatic ring on the hydroxyl carbon. Therefore, the O-H stretching and bending vibrations for saturated and aromatic alcohols fall in the same range. In the spectrum of phenol, the O-H stretch is at 3345 cm^{-1} and the in-plane O-H bend is at 1367 cm^{-1}.

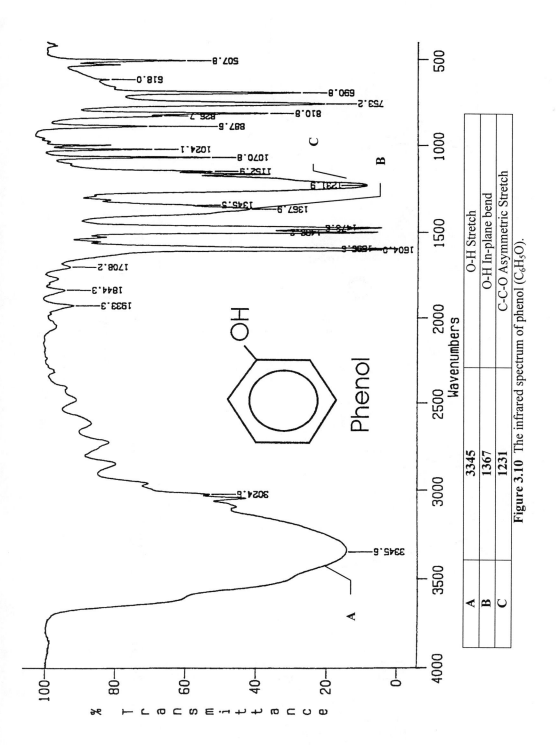

Figure 3.10 The infrared spectrum of phenol (C_6H_5O).

A	3345	O-H Stretch
B	1367	O-H In-plane bend
C	1231	C-C-O Asymmetric Stretch

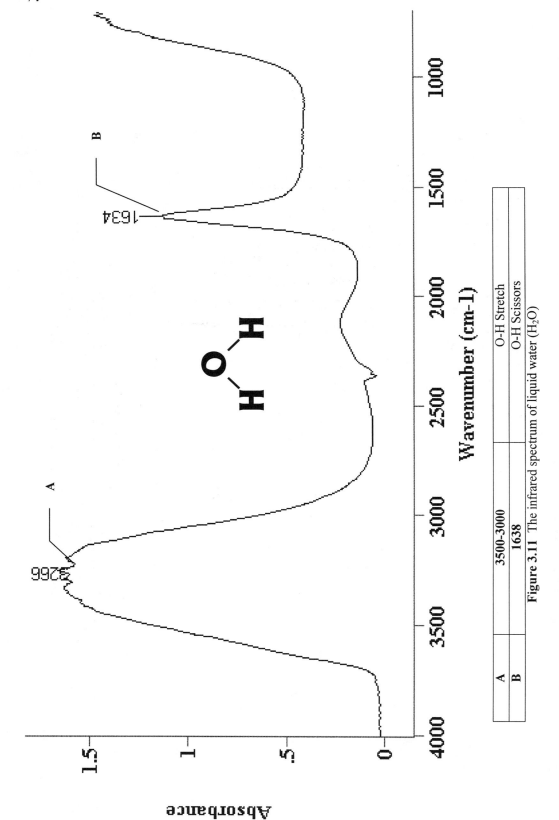

Figure 3.11 The infrared spectrum of liquid water (H_2O)

A	3500-3000	O-H Stretch
B	1638	O-H Scissors

This latter vibration is in the same place as methyl group umbrella modes and benzene ring modes. However, the width of the band at 1367 cm^{-1} indicates it is due to an O-H rather than a C-H or C-C bond. The O-H out-of-plane bending band should appear ~650 cm^{-1}. There is a broad envelope in this region upon which are superimposed many sharp benzene ring bands. This envelope is probably the O-H bending vibration.

Phenols also exhibit an asymmetric C-C-O (or C-O) stretching vibration. In phenols, this band typically shows up between 1260 and 1200 cm^{-1}. In Figure 3.10, this band appears at 1231 cm^{-1}. The phenol and tertiary alcohol C-O stretching band ranges do overlap a little, so it may be difficult to distinguish them if a C-O stretch is near 1200 cm^{-1}. However, for a sample to contain a phenol, it must have an aromatic ring. Therefore, if there is a C-O stretch near 1200 cm^{-1} and no sign of unsaturated C-H stretching, then the molecule must be a tertiary alcohol.

Phenol is also an example of a monosubstituted benzene ring. As a result, it has an unsaturated C-H stretch at 3024 cm^{-1}, ring modes between 1400 and 1600 cm^{-1}, an out-of-plane C-H bend at 753 cm^{-1}, and a ring bend at 690 cm^{-1}. Table 3.1 summarizes the diagnostic infrared bands for the alcohol functional group. It is organized by the substitution pattern on the hydroxyl carbon.

Table 3.1 **The Diagnostic Infrared Bands for Alcohols** (all numbers in cm^{-1})

Subst. Pattern	C-O Stretch	O-H Stretch	O-H Bends
All	-	3350±50	1350±50, 650±50
1°	1075-1000	"	"
2°	1150-1075	"	"
3°	1210-1100	"	"
Phenols	1260-1200	"	"

E. Distinguishing Alcohols from Water

Water is a ubiquitous substance and is found as a common contaminant in many samples. It is particularly common to find water present in alcohol samples since lower molecular weight alcohols are soluble in water. Some alcohols are purchased as water solutions (such as commercial rubbing alcohol, isopropyl alchohol in water). The water molecule is similar to an alcohol since it has two O-H bonds.

Knowledge of the spectrum of water is necessary to distinguish it from alcohols. The infrared spectrum of water, along with its chemical structure, is seen in Figure 3.11. The broad, intense O-H stretch between 3500 and 3000 cm^{-1} is very similar in position and shape to the O-H stretch of an alcohol. In fact, it is impossible to distinguish between an alcohol and water based just on this band position. Fortunately, other sections of the spectrum show differences. The water band at 1630 cm^{-1} is due to the scissoring of the two O-H bonds. This vibration is unique to water. The presence of this band along with an O-H stretch is very strong evidence for the presence of liquid water in a sample. An O-H stretch without a 1630 cm^{-1} band is more likely an alcohol, and other bands due to alcohols discussed above can help confirm this. In a mixture of water and an alcohol, the water bands should be easy to spot, and the presence of C-O stretching and O-H bending bands can help confirm the presence of the alcohol.

II. Ethers

A. Structure and Nomenclature

An *ether* contains a C-O-C linkage and consists of a central oxygen atom with two carbon containing substituents attached to it. The structure of the this functional group is seen in Figure 3.12. The carbon atoms directly attached to the ether are called *ether carbons*. If the two substituents on the oxygen are the same, the ether is said to be symmetrically substituted. If the two substituents are different, the molecule is said to be asymmetrically substituted.

76   The C-O Bond

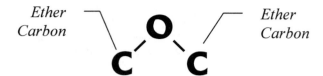

Figure 3.12 The structure of an ether.

If both are saturated, the ether is called a saturated or *dialkyl* ether. If both substituents are aromatic, we have a symmetric aromatic, or diaryl ether. If one substituent is saturated and one is aromatic, we have a mixed, or alkyl aryl ether.

B. The Infrared Spectra of Ethers

Like alcohols, ethers contain a C-O bond and have a strong C-O stretching band. The C-O stretch of an ether involves the C-O-C linkage and is more properly called a "C-O-C asymmetric stretch." (Figure 3.13). We will use the terms C-O stretch and C-O-C asymmetric stretch interchangeably.

Figure 3.13 Ether asymmetric and symmetric C-O-C stretches.

The ether C-O-C asymmetric stretching band is often the most intense between 1300 and 1000 cm^{-1}. There may be a cluster of bands between 1300 and 1000 cm^{-1} in the spectra of ethers due to C-C stretching vibrations. These bands are weak or nonexistent in the spectra of hydrocarbons. However, since the C-O bond is quite polar, the dipole moment change for these vibrations can be large. Different types of ethers can be distinguished by the position and number of C-O-C stretching bands in their spectra.

The C-O-C linkage can undergo a symmetric stretch as seen in Figure 3.13. This band is usually less intense than the asymmetric stretch and is found between 900 and 800 cm^{-1}. This band alone is not always diagnostic for an ether, but should be tracked down so it will not be assigned as being due to any other functional group. Ethers can be distinguished from alcohols because of their lack of O-H stretching and bending bands. Unfortunately, the asymmetric and symmetric C-O-C stretches are the only useful group wavenumbers for ethers. Often an ether is identified after eliminating the presence of other functional groups.

C. Saturated Ethers

The infrared spectrum of a saturated (dialkyl) ether is seen in Figure 3.14. The molecule in this spectrum, diethyl ether, is sometimes called "ether." It is commonly used as a solvent in organic synthesis, and was the first anesthetic used in surgery. The asymmetric C-O-C stretch in this spectrum is clearly visible at 1122 cm^{-1}. For saturated ethers, this vibration shows up between 1150 and 1070 cm^{-1}. Note that there is only one strong band in this region; because the two C-O bonds in a symmetric ether are equivalent. This band appears in the same region as the C-O stretches of some alcohols. However, the absence of an O-H band around 3350 cm^{-1} can confirm that the band in question is the C-O stretch of an ether. not an alcohol. The ether C-O-C symmetric stretch shows weakly in this spectrum at 845 cm^{-1}.

The existence of branch points on an ether carbon causes the C-O stretch region to become more complicated than in unbranched ethers.

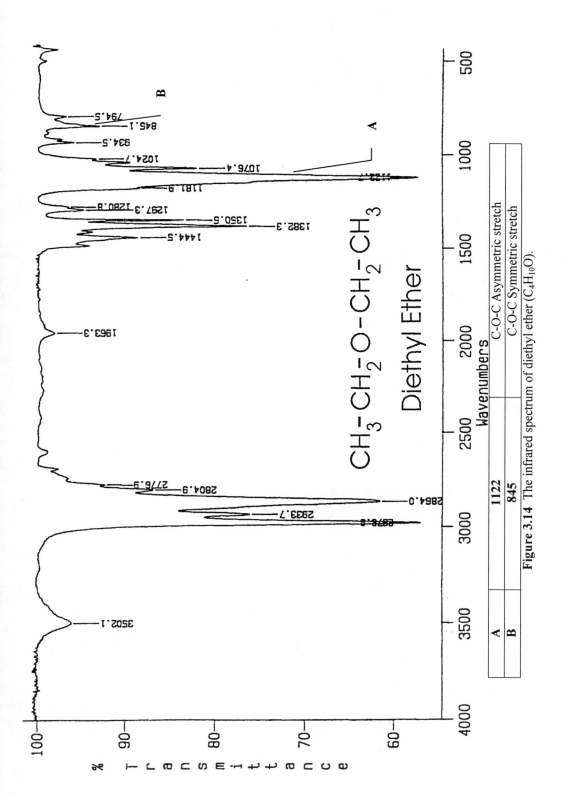

Figure 3.14 The infrared spectrum of diethyl ether ($C_4H_{10}O$).

| A | 1122 | C-O-C Asymmetric stretch |
| B | 845 | C-O-C Symmetric stretch |

This is illustrated in the spectrum of methyl *t*-butyl ether (MTBE) shown in Figure 3.15. MTBE is an important molecule because it has replaced lead as an octane enhancer in gasoline, giving rise to unleaded gasoline. Its spectrum shows intense bands at 1204 and 1085 cm^{-1}. Both of these bands are due to vibrations involving C-O and C-C stretching. Nominally, the band at 1204 cm^{-1} can be assigned as the CH$_3$-C stretch of the *t*-butyl group. The band at 1085 cm^{-1} can be assigned as the C-O-C asymmetric stretch. The important feature of this spectrum is that there are <u>two</u> strong bands between 1210 and 1070 cm^{-1}.

The presence of two (or more) strong bands in this region is an indication that there is branching on the ether carbons. The C-O-C symmetric stretch of MTBE appears at 851 cm^{-1}. Note that the split umbrella mode at 1386 and 1364 has an intensity ratio of 1:2 as expected for a t-butyl group.

D. The Infrared Spectra of Aromatic Ethers

By definition, an aromatic ether has one or two benzene rings directly attached to the ether oxygen. The spectrum of an alkyl aryl ether, anisole (phenyl methyl ether) is shown in Figure 3.16. This is a somewhat complicated spectrum because anisole is a somewhat complicated molecule. However, the bands due to the ether linkage are clearly visible at 1247 and 1040 cm^{-1}. These bands can be nominally assigned as the aromatic ring and saturated C-O stretches, respectively. However, both bands are due to combinations of the asymmetric C-O-C stretch and C-C stretching in the benzene ring. For a mixed ether like anisole, the benzene ring C-O stretch shows up between 1300 and 1200 cm^{-1}, while the saturated C-O stretch appears between 1050 and 1010 cm^{-1}. Thus, a mixed ether can be recognized by a strong band in each of these wavenumber ranges.

In diaryl ethers, there is generally just one strong band of note, the C-O-C asymmetric stretch. This appears between 1300 and 1200 cm^{-1}. Table 3.2 summarizes the group wavenumbers for ethers.

Table 3.2 Group Wavenumbers for Ethers (all numbers in cm^{-1})

Ether Type	Asymmetric C-O-C	Symmetric C-O-C
Saturated, Unbranched	1140-1070 (1 band)	890-820
Saturated, Branched	1210-1070 (2 or more bands)	890-820
Alkyl/Aryl (mixed)	1300-1200 and 1050-1010	-
Aryl	1300-1200	-

E. CH$_3$ and CH$_2$ Groups Attached to an Oxygen

As mentioned in Chapter 2, the methyl and methylene band positions are most reliable when these groups are attached to other carbon atoms. When directly attached to an oxygen atom, the C-H bending and stretching bands can shift position due to electronic effects. The purpose of this section is to discover how to use these band shifts to ascertain what is directly bonded to an oxygen in an alcohol or ether.

For the O-CH$_3$ or *methoxy* group, the methyl asymmetric stretch can fall between 2970 to 2920 cm^{-1}, a much wider range than in a hydrocarbon. The symmetric CH$_3$ stretch of a methoxy group is a better group wavenumber. This band is often sharp, of medium intensity, and falls at 2830±10 cm^{-1} for O-CH$_3$ groups in alcohols and ethers. This is lower than any hydrocarbon C-H stretch, so this band is easy to spot. In the spectrum of methyl t-butyl ether in Figure 3.15, there is a band at 2827 cm^{-1} due to the symmetric CH$_3$ stretch of the methoxy group. In Figure 3.16, the spectrum of anisole shows the same band at 2835 cm^{-1}. Notice that in both these cases the band is separated from the other C-H stretches and is relatively sharp.

Unfortunately, the substitution of an oxygen onto a methyl group throws off the position of one of the best group wavenumbers for a methyl group, the umbrella mode. This band is normally seen at 1375±10 cm^{-1} in the spectrum of a pure hydrocarbon. For a methoxy group, this vibration falls from 1470 to 1440 cm^{-1}, in the same range as other CH$_3$ and CH$_2$ bending vibrations.

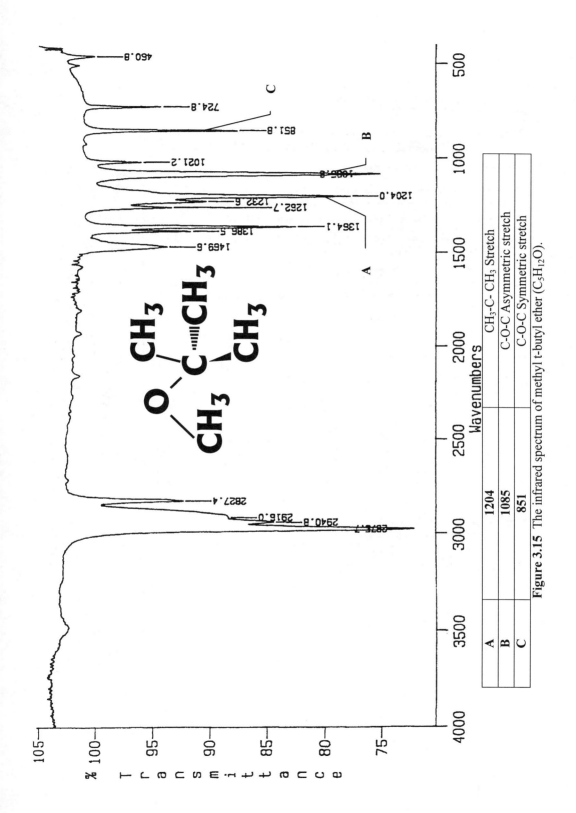

Figure 3.15 The infrared spectrum of methyl t-butyl ether ($C_5H_{12}O$).

A	1204	CH_3-C- CH_3 Stretch
B	1085	C-O-C Asymmetric stretch
C	851	C-O-C Symmetric stretch

80 The C-O Bond

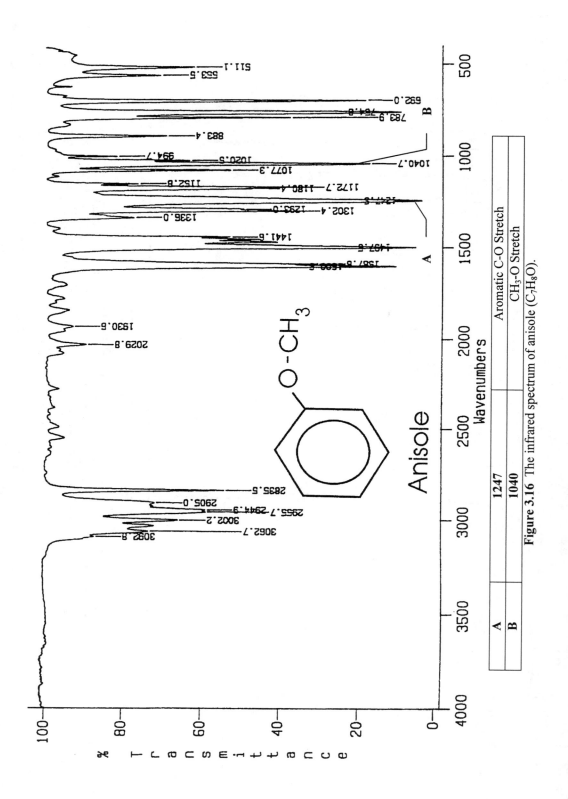

Figure 3.16 The infrared spectrum of anisole (C_7H_8O).

A	1247	Aromatic C-O Stretch
B	1040	CH_3-O Stretch

This renders the umbrella mode of a methoxy group a poor group wavenumber. In the spectrum of MTBE in Figure 3.15, the umbrella modes observed near 1375 cm^{-1} are due to the t-butyl group, not to the methoxy group. In the spectrum of anisole in Figure 3.16, there is no umbrella mode because the only methyl present is in a methoxy group. The umbrella mode in this spectrum is one of the several bands near 1450 cm^{-1}.

It would be nice if the -CH$_2$-O group had a good group wavenumber like the methoxy group. Unfortunately, this is not the case. Oxygen substitution on a methylene causes the asymmetric C-H stretch range to widen to 2955-2920 cm^{-1}, and the symmetric C-H stretch range to widen to 2878-2835 cm^{-1}. These overlap the stretching ranges for CH$_2$s and CH$_3$s in pure hydrocarbons. Sometimes, the symmetric stretch is more intense than the asymmetric stretch in the O-CH$_2$ group.

This is rarely the case in pure hydrocarbons, and should be a flag that there may be a CH$_2$-O linkage present. Table 3.3 summarizes the group wavenumbers for methyl and methylene groups directly attached to an oxygen atom in an alcohol or ether.

Table 3.3 The C-H Stretches and Bends for Hydrocarbons Attached to Oxygen (all numbers in cm^{-1})

Vibration	CH$_3$-O	CH$_2$-O
Asymmetric C-H stretch	2970-2920	2955-2920
Symmetric C-H stretch	2830±10	2878-2835
CH$_3$ Umbrella mode	1470-1440	-

Bibliography

L.J. Bellamy, *The Infrared Spectra of Complex Molecules, Third Edition*, Wiley, New York, 1975.

R. Silverstein, G. Bassler, T. Morrill, *Spectrometric Identification of Organic Compounds, Fourth Edition*, Wiley, New York, 1981.

D. Lin-Vien, N. Colthup, W. Fately, J. Grasselli, *Infrared and Raman Characteristic Frequencies of Organic Molecules*, Academic Press, Boston, 1991.

N. Colthup, L. Daly, S. Wiberley, *Introduction to Infrared and Raman Spectroscopy*, Academic Press, New York, 1990.

F. Miller, *Lecture Notes for Bowdoin College Infrared Short Course*, Brunswick, Maine, 1992.

B.C. Smith, *Fundamentals of Fourier Transform Infrared Spectroscopy*, CRC Press, Boca Raton, 1996.

H. Mantsch and D. Chapman, Eds., *Infrared Spectroscopy of Biomolecules*, Wiley, New York, 1996.

A. Streitweiser and C. Heathcock, *Introduction to Organic Chemistry*, Macmillan, New York, 1976.

S. Budavari, Editor, *The Merck Index, 11th Edition*, Merck & Co., Rahway, NJ, 1989.

R.C. Weast, Editor, *The CRC Handbook of Chemistry and Physics*, CRC Press, Boca Raton, 1987.

J. Dean, Editor, *Lange's Handbook of Chemistry*, McGraw-Hill, New York, 1979.

Problem Spectra

The following spectra are presented as an exercise for the reader. Using the techniques and knowledge obtained in this chapter, do your best to predict the complete molecular structure of each sample from its infrared spectrum. The exact positions of prominent peaks are contained in the table accompanying each spectrum. However, these peaks may or may not be useful in the final determination of a structure. The physical state of the sample is also stated (solid or liquid) along with the sampling method used.

All problem set spectra were obtained on modern day FTIR instruments at 4 or 8 cm^{-1} resolution and are of pure materials. It is much easier to learn the basics of interpretation by looking at the simple spectra of pure compounds than to tackle the much more complex problem of analyzing mixture spectra. Please realize that the determination of a complete molecular structure in each problem may not be possible. However, by analyzing each of these spectra methodically and patiently, you will be able to learn a lot from each of them. The correct structure and a discussion of the interpretation of each problem are included in an appendix at the end of the book. Good Luck!

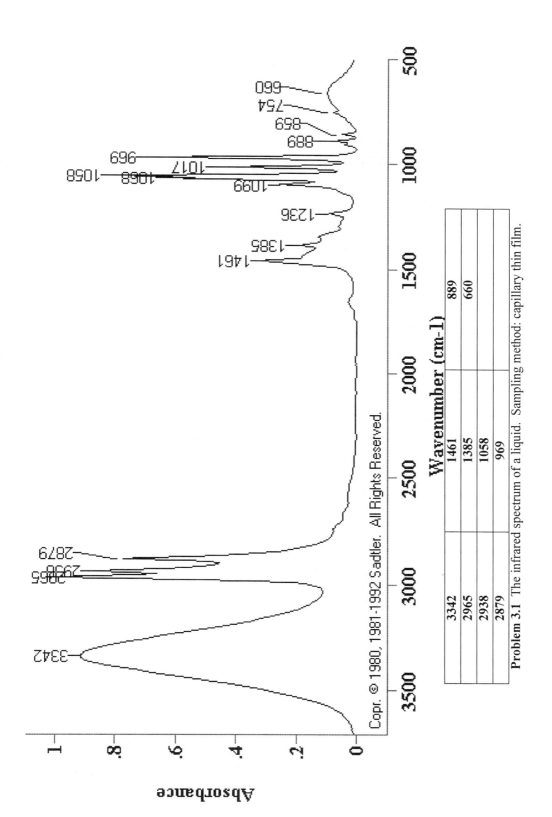

Problem 3.1 The infrared spectrum of a liquid. Sampling method: capillary thin film.

3342	1461	889
2965	1385	660
2938	1058	
2879	969	

84 The C-O Bond

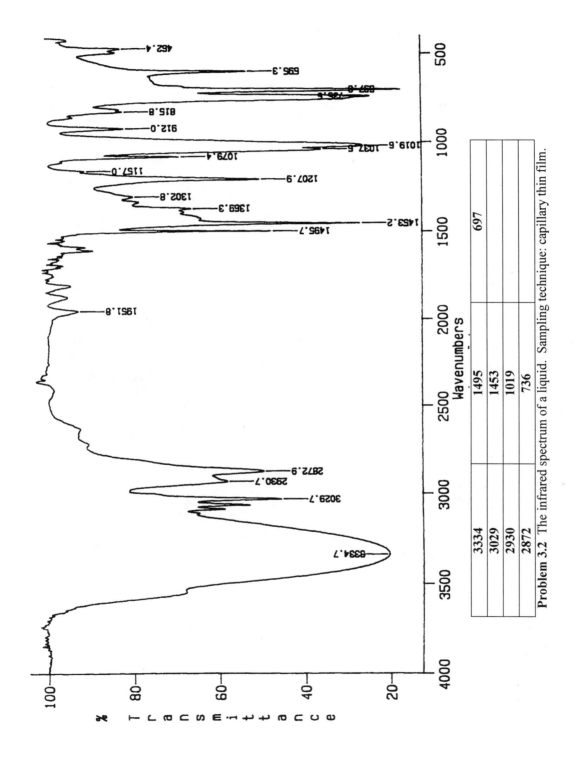

3334	1495	697
3029	1453	
2930	1019	
2872	736	

Problem 3.2 The infrared spectrum of a liquid. Sampling technique: capillary thin film.

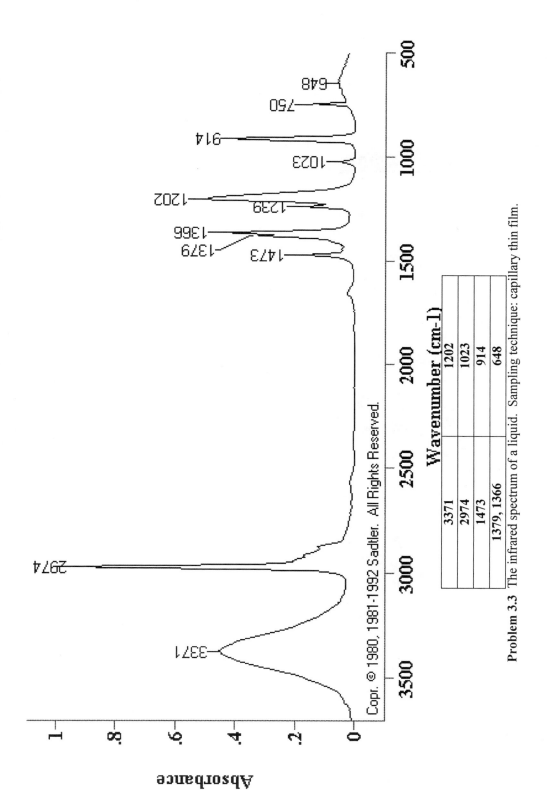

Problem 3.3 The infrared spectrum of a liquid. Sampling technique: capillary thin film.

Wavenumber (cm-1)	
3371	1202
2974	1023
1473	914
1379, 1366	648

86 The C-O Bond

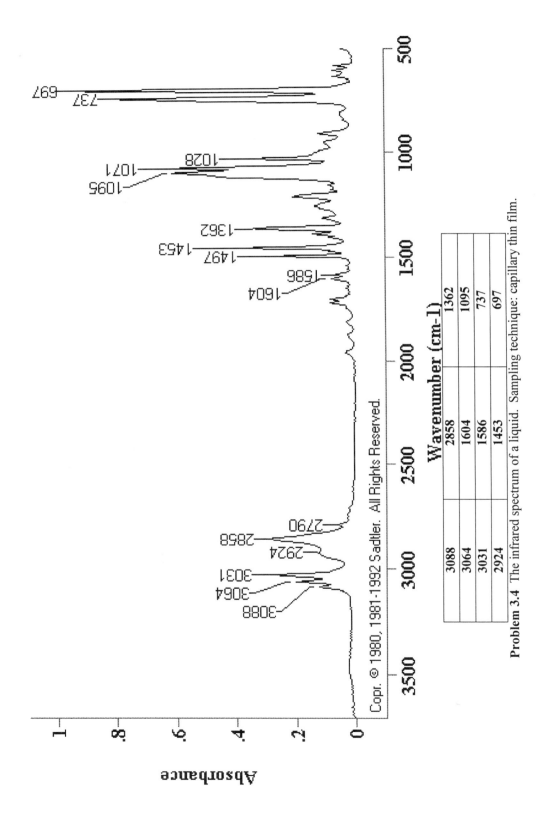

3088	1362
3064	1095
3031	737
2924	697
2858	
1604	
1586	
1453	

Problem 3.4 The infrared spectrum of a liquid. Sampling technique: capillary thin film.

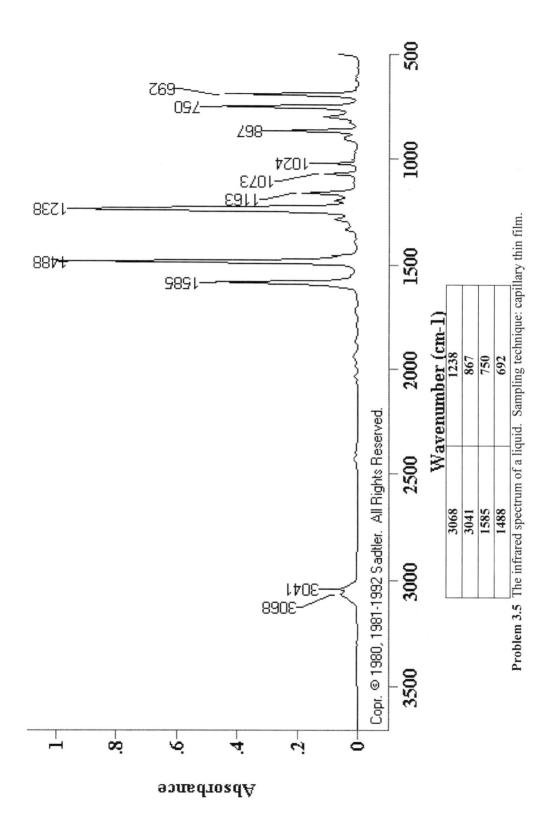

Problem 3.5 The infrared spectrum of a liquid. Sampling technique: capillary thin film.

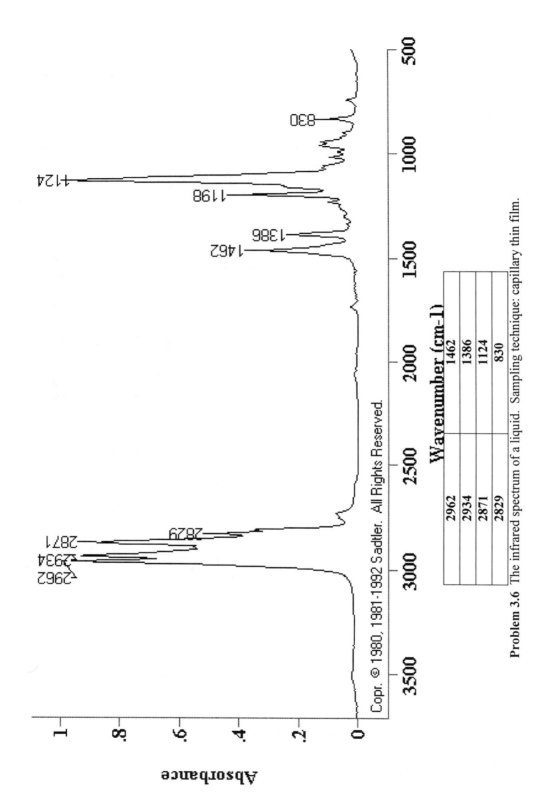

Problem 3.6 The infrared spectrum of a liquid. Sampling technique: capillary thin film.

Wavenumber (cm-1)	
2962	1462
2934	1386
2871	1124
2829	830

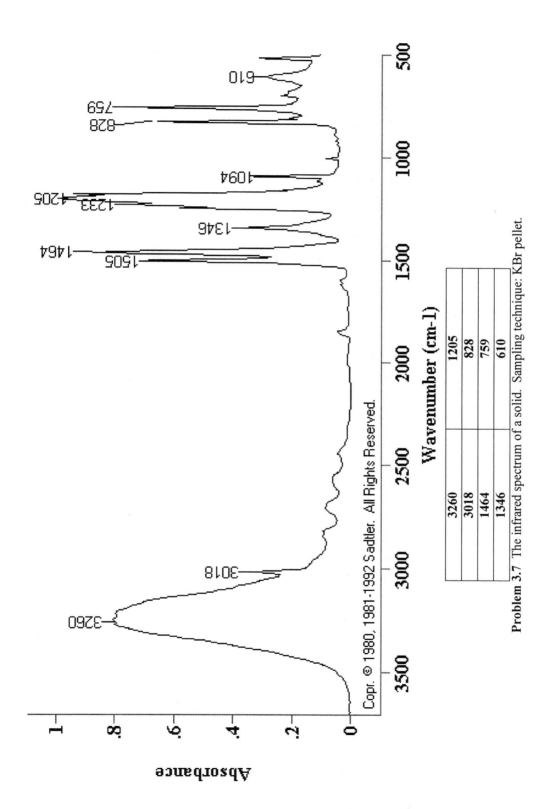

Problem 3.7 The infrared spectrum of a solid. Sampling technique: KBr pellet.

90   The C-O Bond

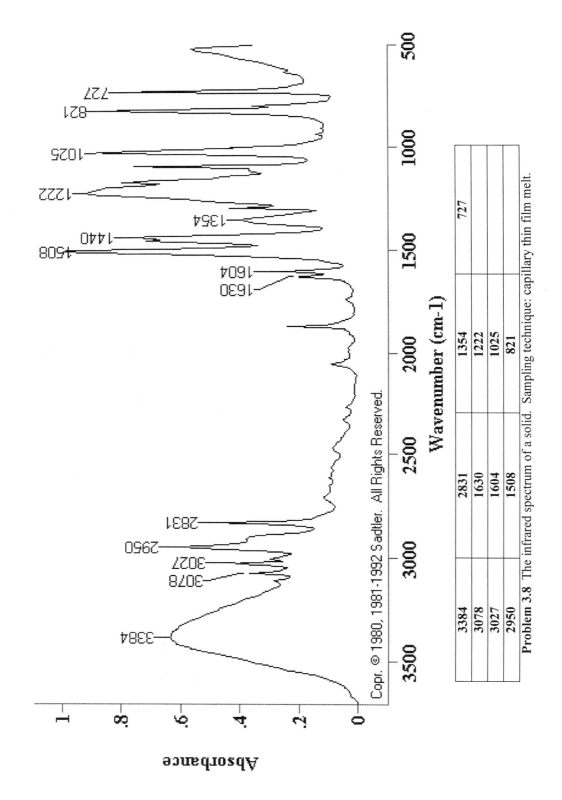

3384	2831	1354	727
3078	1630	1222	
3027	1604	1025	
2950	1508	821	

Problem 3.8 The infrared spectrum of a solid. Sampling technique: capillary thin film melt.

Chapter 4

The Carbonyl Functional Group

I. Introduction

The carbonyl functional group consists of a carbon doubly bonded to an oxygen and has the chemical structure C=O. The structural framework of the carbonyl group is shown in Figure 4.1. The carbon bonded to the oxygen is called the *carbonyl carbon*. Many different types of functional groups can be constructed by attaching different substituents to the carbonyl carbon. For example, ketones, aldehydes, esters, and carboxylic acids all contain C=O bonds, but differ in the nature of the carbonyl carbon substituents. The structure, nomenclature, and spectroscopy of many of these functional groups will be discussed in this chapter.

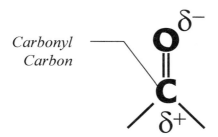

Figure 4.1 The structure of the carbonyl functional group. Note the position of the carbonyl carbon. The δ^+ and δ^- symbols represent partial charges.

Because of the electronegativity differences between the carbon and oxygen atoms in a carbonyl group, the C=O bond has a large dipole moment. This is shown in Figure 4.1 by the partial positive charge on the carbon atom and the partial negative charge on the oxygen atom. Recall from Chapter 1 that the change in dipole moment with respect to bond distance for a vibration determines the absolute intensity of an infrared band. Since the carbonyl group has a large dipole moment, infrared bands due to the C=O group are intense.

The spectral signature of the C=O group is the C=O stretching vibration, which appears as an intense band between 1800 and 1600 cm^{-1}. This vibration is illustrated in Figure 4.2.

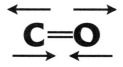

Figure 4.2 The lengthening and contraction of the C=O bond during the stretching vibration of the carbonyl functional group.

The carbonyl stretching vibration involves both the carbon and oxygen atom moving at the same time and is a perfect example of a group wavenumber. It is intense, shows up in a unique spectral region, and appears in a reasonably narrow wavenumber range. Most intense bands between 1800 and 1600 cm^{-1} are C=O stretches.

There are some influences that cause the position of the C=O stretch band to move, such as changes in state and hydrogen bonding. The most relevant of these influences is the impact of the substitution of a benzene ring on the carbonyl carbon. For acetone (see structure in Figure 4.5), which has two saturated substituents on the carbonyl carbon, the C=O stretch falls at 1715 cm^{-1}. If one of the methyl groups in acetone is replaced with a benzene ring, the molecule is called acetophenone (see structure in Figure 4.6), and the C=O stretch shifts to 1683 cm^{-1}. This wavenumber shift is an example of an electronic effect and is caused by a phenomenon called *conjugation*.

To understand conjugation, we need to think about the electrons in carbonyl groups and aromatic rings, and what orbitals they occupy. As explained in Chapter 2, some of the electrons involved in C-C bonding in a benzene ring circulate inside the ring in orbitals that are perpendicular to the plane of the ring. These are called "p-type" orbitals. The double bond of a carbonyl group also contains electrons in p orbitals that are perpendicular to the C=O bond axis. The carbon atom in a benzene ring attached to a carbonyl carbon is in the same plane as the carbonyl carbon. The p orbitals on the two carbons can overlap, allowing electron density from the benzene ring to "spill over" into the C=O bond, and vice versa. This reduces the force constant of the C=O bond, causing the position of the C=O stretching band to shift to lower wavenumber. In general, a single aromatic ring attached to a carbonyl carbon causes the C=O stretch to shift by 20 to 30 cm^{-1} compared to a carbonyl carbon with saturated substituents. Two aromatic rings attached to a carbonyl carbon can cause a C=O stretching shift of greater than 30 cm^{-1}. For most of the functional groups discussed in this chapter, two different C=O stretching wavenumber ranges will be defined. One for saturated (alkyl) substituents attached to the carbonyl carbon, the second for aryl substituents attached to the carbonyl carbon.

In the discussions that follow, the position of the C=O stretching vibration will not tell us what is attached to the carbonyl carbon. For example, the C=O stretching vibrations of ketones, aldehydes, esters, and carboxylic acids all appear between 1750 and 1700 cm^{-1}. We will have to depend on other bands in the infrared spectrum to determine what substituents are attached to the carbonyl carbon.

II. Ketones

A. Structure and Nomenclature

A ketone is any molecule where both substituents on the carbonyl carbon are hydrocarbons. By definition, a ketone has two carbon atoms attached to the carbonyl carbon. The skeletal framework of a ketone is shown in Figure 4.3.

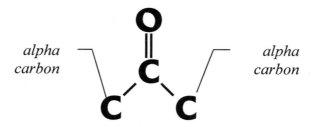

Figure 4.3 The skeletal framework of a ketone. Note that there are two carbon atoms attached to the carbonyl carbon; these carbon atoms are called "alpha carbons."

The atoms directly attached to the carbonyl carbon are called alpha carbons. If both substituents attached to the carbonyl carbon are saturated, then the molecule is called a saturated or *alkyl ketone*. If both substituents are benzene rings, the molecule is an *aryl ketone*. If one substituent is saturated and the other is aromatic, then the molecule is a mixed or alkyl/aryl ketone. Ketones can also be classified by symmetry. If the two hydrocarbon substituents are identical, the ketone is symmetric; if the substituents are different, the ketone is asymmetric.

B. The Infrared Spectra of Ketones

Acetone, or dimethyl ketone, is the simplest ketone. It consists of two methyl groups attached to a carbonyl carbon and is classified as a saturated ketone. The infrared spectrum of acetone is shown in Figure 4.5. The carbonyl stretch appears at 1715 cm^{-1} and, for saturated ketones in general this vibration occurs at 1715±10 cm^{-1}. The position of this band is not sufficient to determine whether a molecule is a ketone. Other parts of the spectrum must be consulted to accomplish this task. The second important group wavenumber for ketones is the "C-C-C" stretch, which involves the asymmetric stretching of the two alpha carbon C-C bonds. An illustration of this vibration is shown in Figure 4.4.

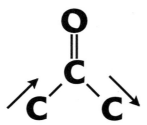

Figure 4.4 The C-C-C stretch of the ketone functional group.

For the acetone molecule, this vibration involves one methyl group moving toward the carbonyl carbon while the second methyl group moves away. In hydrocarbons, C-C stretching vibrations are weak because of the small (or nonexistent) dipole moment of these bonds. However, the carbonyl carbon has a partial positive charge on it, and any C-C bond that includes the carbonyl carbon will have a significant dipole moment. The C-C-C stretch in the spectrum of acetone is seen at 1222 cm^{-1}. For saturated ketones, this vibration appears from 1230 to 1100 cm^{-1}. Many other functional groups, such as C-O bonds, have bands in this region. However, the combination of a C=O stretch and a C-C-C stretch are strong evidence for the existence of a ketone in a sample.

An alkyl/aryl ketone can be distinguished from a saturated ketone based on the position of the C=O and C-C-C stretches. The spectrum of an alkyl/aryl ketone, acetophenone, is shown in Figure 4.6. The C=O stretching wavenumber is lowered by conjugation with the benzene ring and appears at 1684 cm^{-1}. For mixed ketones, the C=O stretch appears from 1700 to 1670 cm^{-1}. The C-C-C stretch of acetophenone is at 1266 cm^{-1}, and generally appears for mixed ketones from 1300 to 1230 cm^{-1}. For a diaryl ketone, conjugation has an even greater impact on the position of the C=O stretch than in a mixed ketone. The C=O stretch of diphenylketone (benzophenone) occurs at 1652 cm^{-1}. For most diaryl ketones, this band appears between 1680 and 1600 cm^{-1}. The C-C-C stretch of benzophenone appears at 1264 cm^{-1} and, for diaryl ketones, is generally found between 1300 and 1230 cm^{-1}. This is the same range as for mixed ketones. Thus, the 1710 cm^{-1} point divides saturated from mixed and aromatic C=O stretches, and the 1230 cm^{-1} point divides saturated from mixed and aromatic C-C-C stretches. Unfortunately, there are no bands in the infrared spectrum of a ketone that disclose whether it is symmetrically or asymmetrically substituted.

94 The Carbonyl Functional Group

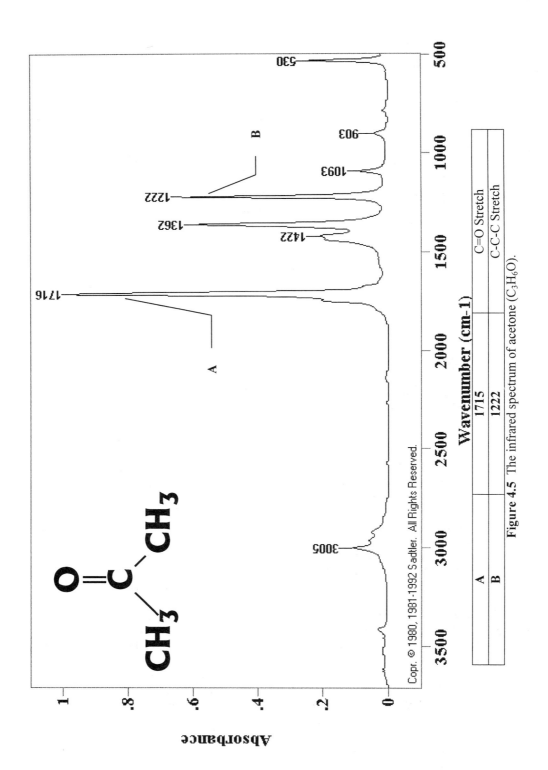

Figure 4.5 The infrared spectrum of acetone (C_3H_6O).

A	1715	C=O Stretch
B	1222	C-C-C Stretch

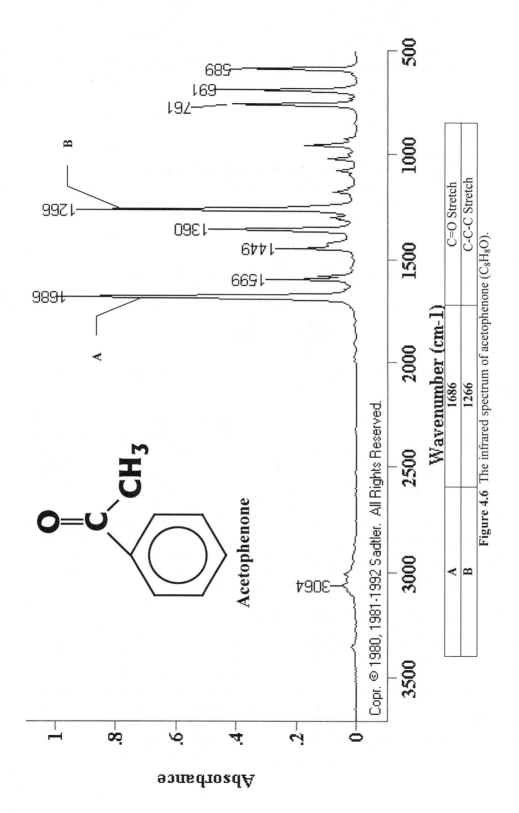

Figure 4.6 The infrared spectrum of acetophenone (C_8H_8O).

A	1686	C=O Stretch
B	1266	C-C-C Stretch

Table 4.1 The Group Wavenumbers of Ketones (all numbers in cm^{-1})

Vibration	Wavenumber Range
Saturated C=O stretch	1715±10
Aromatic C=O stretch	1700-1640
Saturated C-C-C stretch	1230-1100
Aryl/Alkyl and Diaryl C-C-C stretch	1300-1230

III. Aldehydes

A. Structure and Nomenclature

An aldehyde consists of a carbon atom and a hydrogen atom attached to a carbonyl carbon. The skeletal framework of an aldehyde is seen in Figure 4.7.

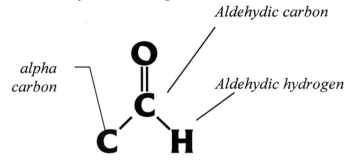

Figure 4.7 The skeletal framework of an aldehyde. Note the locations of the aldehydic hydrogen, aldehydic carbon, and alpha carbon.

The carbonyl group and the carbon and hydrogen attached to it are all in the same plane. The carbonyl carbon in an aldehyde is called the *aldehydic carbon*, and the hydrogen attached to the carbonyl carbon is called the *aldehydic hydrogen*. The carbon directly attached to the carbonyl group is called the "alpha carbon," which will be abbreviated "α carbon." The aldehydic hydrogen is in a unique position, being directly attached to the carbonyl carbon and separated from the rest of the aldehyde molecule by the carbonyl group. The aldehydic hydrogen is responsible for many of the unique spectral properties of aldehydes.

Aldehydes can be divided into saturated and aromatic aldehydes. A saturated aldehyde has a saturated hydrocarbon chain attached to the aldehydic carbon. An aromatic aldehyde has an aromatic ring directly attached to the aldehydic carbon. The carbonyl group in an aromatic aldehyde conjugates with the adjacent benzene ring, affecting the force constants of the C=O bond and causing the C=O stretch to shift to lower wavenumbers as in ketones.

B. Infrared Spectroscopy of Aldehydes

One of the most important group wavenumbers for aldehydes is the C-H stretching vibration of the aldehydic hydrogen. The aldehydic C-H stretch occurs between 2850 and 2700 cm^{-1}, lower than most other C-H stretches. It is easy to distinguish an aldehydic C-H stretch from that of a methyl or methylene group by its low wavenumber. Often the aldehydic C-H stretch may be accompanied by a second band of almost equal intensity in the 2850 to 2700 cm^{-1} range. This second band is the first overtone of the aldehydic C-H bending vibration (which appears near 1390 cm^{-1} as discussed below). This overtone band appears in the spectrum by stealing intensity from the C-H stretching band via Fermi resonance (see Chapter 1). A pair of bands between 2850 and 2700 cm^{-1} is strong evidence for the presence of an aldehyde in a sample, but only <u>some</u> aldehydes exhibit both bands.

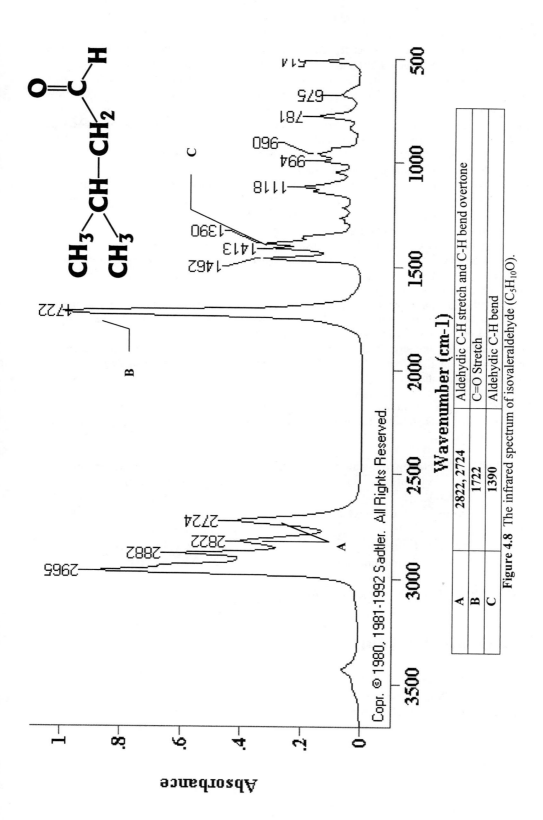

Figure 4.8 The infrared spectrum of isovaleraldehyde ($C_5H_{10}O$).

A	2822, 2724	Aldehydic C-H stretch and C-H bend overtone
B	1722	C=O Stretch
C	1390	Aldehydic C-H bend

The infrared spectrum of isovaleraldehyde, seen in Figure 4.8, has peaks at 2824 and 2724 cm^{-1} due to the C-H stretching vibration and the overtone of the C-H bending vibration.

The position of the aldehydic C-H stretch in saturated aldehydes is somewhat sensitive to branching on the α carbon. If the α carbon is unbranched (a CH$_3$ or CH$_2$ group), the aldehydic C-H stretch appears between 2730 and 2715 cm^{-1}. If the α carbon is branched (a methine or quaternary carbon), the aldehydic C-H stretch appears between 2715 and 2000 cm^{-1}. If there is a pair of bands between 2850 and 2700 cm^{-1} due to Fermi resonance, use the lower wavenumber of the two bands to determine the branching on the α carbon. The spectrum of isovaleraldehyde in Figure 4.8 is consistent with these rules. It is a saturated aldehyde where the alpha carbon is a methylene group. The lower wavenumber of the two bands in the 2850 to 2700 cm^{-1} range is at 2724, as would be expected for a saturated, unbranched aldehyde.

The aldehydic C-H bending vibration appears at 1390±10 cm^{-1} with medium to weak intensity. The carbonyl group and the two substituents in an aldehyde all lie in the same plane. This C-H bending vibration is an in-plane vibration, meaning the hydrogen does not leave the plane of the aldehyde group during the vibration. Many other bands appear around 1400 cm^{-1}, so it is not always possible to distinguish the aldehydic C-H bend.

Like most carbonyl containing functional groups, aldehydes have a strong C=O stretching band that is often the most intense band in the spectrum. This vibration shows up in saturated aldehydes at 1730±10 cm^{-1} and is seen in Figure 4.8 at 1720 cm^{-1}. Because of conjugation, the C=O stretch of aromatic aldehydes shows up between 1710 and 1685 cm^{-1}. The large dipole moment of the carbonyl group causes the C-C bond in an aldehyde to have an appreciable dipole moment. The stretching of this bond, which we will call an aldehyde "C-C stretch," gives rise to one to several medium intensity bands between about 1400 and 1100 cm^{-1}. For aromatic aldehydes, the stretching of the alpha carbon/carbonyl carbon bond gives a medium intensity band from 1210 to 1160 cm^{-1}. Because many other functional groups absorb between 1400 and 1100 cm^{-1}, this band is not a good group wavenumber, but it can be used as secondary evidence for the presence of an aldehyde.

In conclusion, the best indicators of an aldehyde are a carbonyl stretch in combination with a low wavenumber C-H stretch. Table 4.2 summarizes the group wavenumbers for the aldehyde functional group.

Table 4.2 The Group Wavenumbers of the Aldehyde Functional Group (in cm^{-1})

Vibration	Wavenumber Range
Saturated C=O Stretch	1730±10
Aromatic C=O Stretch	1710-1685
C-H Bend	1390±10
Aldehydic C-H Stretch, general	2850-2700 (1 or 2 bands)
Aldehydic C-H Stretch, unbranched α carbon	2730-2715 (1 band)
Aldehydic C-H Stretch, branched α carbon	2715-2700 (1 band)

IV. Carboxylic Acids and Their Derivatives

A. Carboxylic Acids
1. Structure and Nomenclature
The skeletal framework of a carboxylic acid is shown in Figure 4.9.

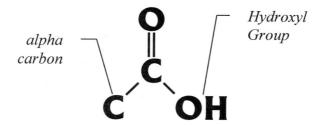

Figure 4.9 The skeletal framework of a carboxylic acid. Note the location of the alpha carbon.

Note that by definition a carboxylic acid consists of a carbonyl group with an O-H (hydroxyl) group attached on one side, and a carbon atom (the alpha or α carbon) attached on the other side. If the alpha carbon is saturated, then the molecule is called a saturated acid. If the alpha carbon is part of a benzene ring, the molecule is called an aromatic acid.

Carboxylic acids are called "acids" because the O-H group can lose its hydrogen and ionize like other acids. Carboxylic acids are not as strong as hydrochloric or nitric acids, but do undergo the classic reactions of acids, such as reacting with bases to form a salt and water.

Carboxylic acids are similar to alcohols and water in that they all contain the O-H group. This structural unit is responsible for hydrogen bonding, and carboxylic acids are an extreme example of hydrogen bonding. The hydrogen bonds in carboxylic acids are so strong that the acid molecules are bound together in dimers when in the solid and liquid state. Figure 4.10 shows an example of a carboxylic acid dimer. Note how the O-H of one molecule hydrogen bonds to the C=O of a second molecule, and vice versa.

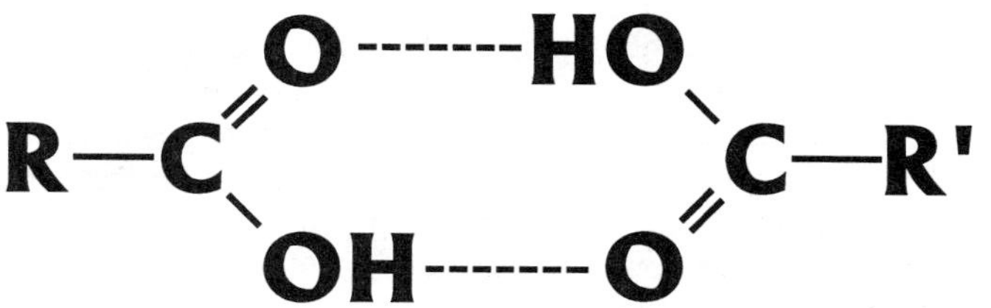

Figure 4.10 The carboxylic acid dimer form, typically found in solid and liquid samples of these molecules. The dashed lines represent the hydrogen bonds between two different carboxylic acid molecules.

Carboxylic acids are found only in monomer form in the vapor phase and in dilute solution. In most samples, the infrared spectrum of the dimer is measured. The infrared bands discussed below for carboxylic acids all refer to the dimer form.

2. The Infrared Spectra of Carboxylic Acids

Since a carboxylic acid contains a carbonyl group and a hydroxyl group, it shows bands characteristic of both these structural units. Recall from the discussion in Chapter 1 that any molecule that undergoes hydrogen bonding has broad infrared bands, and that bands due to the O-H group are particularly broad. All of this is true for carboxylic acids.

The overriding spectral feature of a carboxylic acid is the broad, intense O-H stretching band typically found from 3500 to 2500 cm^{-1}, and often centered around 3000 cm^{-1}. In Figure 4.11, the full width at half height of this band is 800 cm^{-1}! No other functional group has such a broad, intense band at high wavenumber. This band almost, by itself, tells you that a sample contains a carboxylic acid. In the spectrum of 2-methylbutyric acid in Figure 4.11, the sharper C-H stretching vibrations are superimposed upon the broad O-H stretching band. This is common in the spectra of acids. However, sometimes the O-H stretch masks the C-H stretching bands. Note that on the low wavenumber side of the O-H stretching band in Figure 4.11, between 2500 and 2800 cm^{-1} there are some broad features of medium intensity. These are overtone and combination bands of lower wavenumber C-C stretching and C-H bending vibrations that appear in the spectrum by stealing intensity via Fermi resonance (see Chapter 1). These features are quite common in the spectra of carboxylic acids.

Like the O-H group in an alcohol, the O-H group in a carboxylic acid can bend in and out of the plane defined by the acid moiety. The in-plane O-H bending band is found from 1440 to 1395 cm^{-1}. In the spectrum in Figure 4.11, this band is found at 1418 cm^{-1}. The in-plane O-H bending band appears in the same region as C-H bending vibrations such as the methyl group umbrella mode. However, O-H bending bands are usually broader than C-H bending bands, making them easy to distinguish. For example, in Figure 4.11 the O-H in-plane bending band at 1418 cm^{-1} is broader than the CH$_3$ umbrella mode at 1384 cm^{-1}.

The out-of-plane O-H bending band is almost as diagnostic for the presence of carboxylic acids as the O-H stretch. Its low wavenumber, medium intensity, and breadth make it unique. This band is found in Figure 4.11 at 942 cm^{-1}, and is generally found between 960 and 900 cm^{-1}. Carboxylic acids contain a C-O bond, and the C-O stretch of carboxylic acids appears between 1320 and 1210 cm^{-1}. In the spectrum of 2-methylbutyric acid in Figure 4.11, the C-O stretch is seen at 1230 cm^{-1}. This band is often times the most intense in this region of the spectrum.

The last of the many useful group wavenumbers for carboxylic acids is the C=O stretch. This vibration is seen from 1730 to 1700 cm^{-1} in saturated acids and from 1710 to 1680 in aromatic acids. The aromatic acid C=O stretch falls at lower wavenumber than the saturated C=O stretch due to conjugation of the carbonyl group with the benzene ring. The C=O stretch of 2-methylbutyric acid falls at 1709 cm^{-1} as seen in Figure 4.11. The best way to distinguish between saturated and aromatic acids is by the position of the C=O stretch.

The C=O stretch of benzoic acid is seen in Figure 4.12 at 1685 cm^{-1}. As would be expected for any carboxylic acid, the broad O-H stretch of benzoic acid is seen between 3500 and 2000 cm^{-1}, and there are overtone and combination bands superimposed upon it around 2600 cm^{-1}. The O-H bending bands stand out because of their width, and are found at 1423 and 934 cm^{-1}. The C-O stretch is seen at 1292 cm^{-1}. The spectrum of benzoic acid is interesting because it has broad bands due to the carboxylic acid and narrow bands due to the benzene ring. Although there are many bands in this spectrum, by remembering which functional groups have wide and narrow bands, this spectrum can be correctly assigned.

In summary, the C=O stretch of carboxylic acids is in the same region as the C=O stretch of ketones, aldehydes, and esters. The C=O stretching band by is not sufficient evidence to identify a carboxylic acid. Other bands, such as the O-H stretch and out-of-plane O-H bend must be found to confirm the assignment. Table 4.3 summarizes the group wavenumbers for carboxylic acids.

Table 4.3 The Group Wavenumbers of Carboxylic Acids (all numbers in cm^{-1})

Vibration	Wavenumber Range
Saturated C=O Stretch	1730-1700
Aromatic C=O Stretch	1710-1680
C-O Stretch	1320-1210
O-H Stretch	3500-2500 (broad and intense)
O-H In-plane bend	1440-1395
O-H Out-of-plane bend	960-900

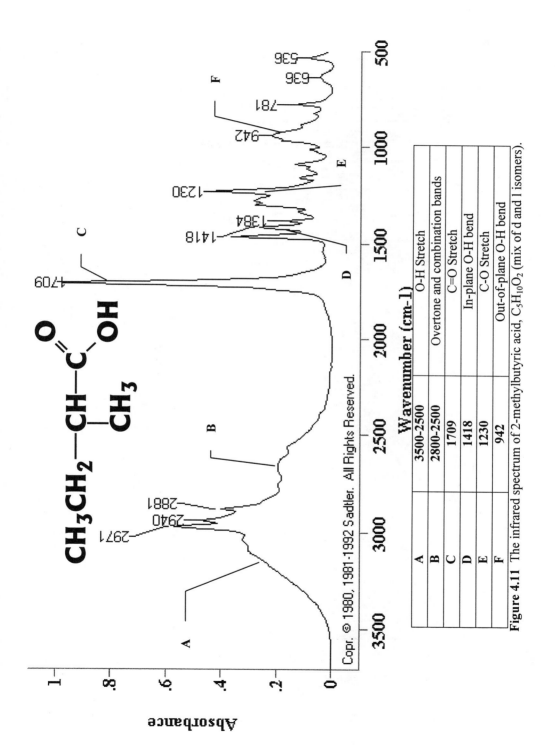

Figure 4.11 The infrared spectrum of 2-methylbutyric acid, $C_5H_{10}O_2$ (mix of d and l isomers).

	Wavenumber (cm-1)	
A	3500-2500	O-H Stretch
B	2800-2500	Overtone and combination bands
C	1709	C=O Stretch
D	1418	In-plane O-H bend
E	1230	C-O Stretch
F	942	Out-of-plane O-H bend

102   The Carbonyl Functional Group

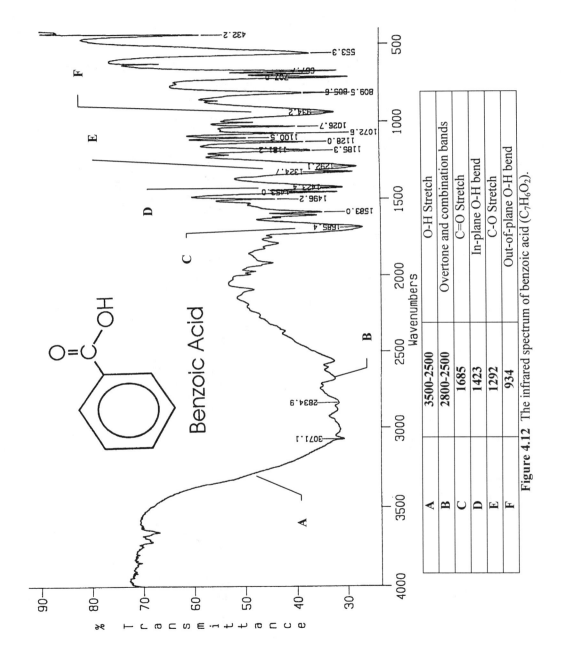

Figure 4.12 The infrared spectrum of benzoic acid ($C_7H_6O_2$).

A	3500–2500	O-H Stretch
B	2800–2500	Overtone and combination bands
C	1685	C=O Stretch
D	1423	In-plane O-H bend
E	1292	C-O Stretch
F	934	Out-of-plane O-H bend

B. Carboxylic Acid Salts: Carboxylates
1. Structure and Nomenclature
A *salt* is the product of a reaction between an acid and a base. When a carboxylic acid reacts with a base, the products are a *carboxylate* salt and a molecule of water. The reaction that produces a carboxylate is shown in Figure 4.13.

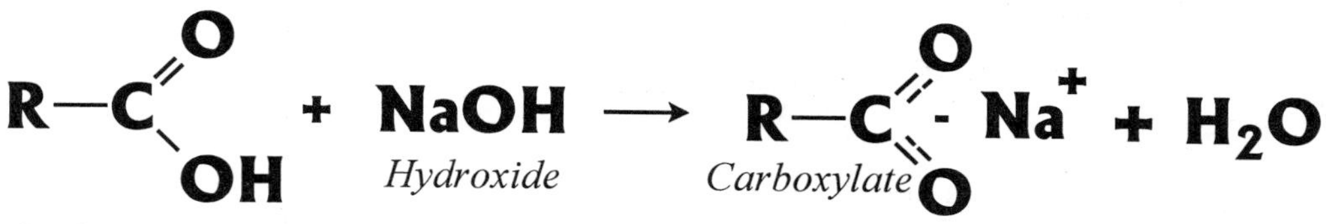

$$R-C(=O)(OH) + NaOH \rightarrow R-C(\cdots O)(\cdots O)^- Na^+ + H_2O$$

Carboxylic Acid   Hydroxide   Carboxylate

Figure 4.13 The reaction between a carboxylic acid and a hydroxide to produce a carboxylate.

Note that the O-H of the acid is not present in the carboxylate, having reacted to form part of the water molecule that is a byproduct of the reaction.

The bonding arrangement in a carboxylate is rather unusual. The two oxygens in the carboxylate coordinate to the metal atom from the hydroxide. The two carbon-oxygen bonds contain only 3 electrons between them. The two bonds can be thought of as "bonds and a half." The two bonds are equivalent, and the third electron can be thought of as being shared by the two C-O bonds. The dashed lines in Figure 4.13 represent this third electron. Carboxylates are named after the acid from which they are derived and the metal atom to which the oxygens are coordinated. For example, a reaction between sodium hydroxide and acetic acid would produce sodium acetate. The main ingredient in soap, sodium stearate, is a carboxylate produced by the reaction of sodium hydroxide and stearic acid. It is the cleansing power of carboxylates, and there common use in soaps and detergents that makes them economically important molecules.

2. The Infrared Spectra of Carboxylates
Although carboxylates are derived from carboxylic acids, the lack of an O-H group makes their spectra radically different. The spectrum of a carboxylate, zinc stearate, is seen in Figure 4.14. Zinc stearate is made by reacting stearic acid with zinc hydroxide. Note the lack of broad O-H stretching and bending bands. The two "bond and a half" linkages in a carboxylate can stretch asymmetrically and symmetrically. We will refer to these vibrations as "CO_2" stretches since the carbon and both oxygens are involved. The CO_2 stretches produce two strong, unique infrared absorbance bands. The intensity of these bands is due to the large dipole moment of the carbon-oxygen bonds. The asymmetric CO_2 stretch generally falls between 1650 and 1540 cm^{-1}, and the symmetric CO_2 stretch is found between 1450 and 1360 cm^{-1}. In Figure 4.14 the CO_2 stretches of zinc stearate are found at 1537 and 1398 cm^{-1}. The asymmetric stretch is usually the more intense of these two bands. Both of these bands must be present in a spectrum for there to be a carboxylate in a sample. Many other functional groups absorb in the same wavenumber ranges as carboxylates. However, the intensity of these bands, and the fact that there must be two of them in specific wavenumber ranges, makes carboxylates relatively easy to identify from their infrared spectra. Table 4.4 summarizes the CO_2 stretching information for carboxylates.

Table 4.4 The Group Wavenumbers of Carboxylates (all numbers in cm^{-1})

Vibration	Wavenumber Range
Asymmetric CO_2 stretch	1650-1540
Symmetric CO_2 stretch	1450-1360

104   The Carbonyl Functional Group

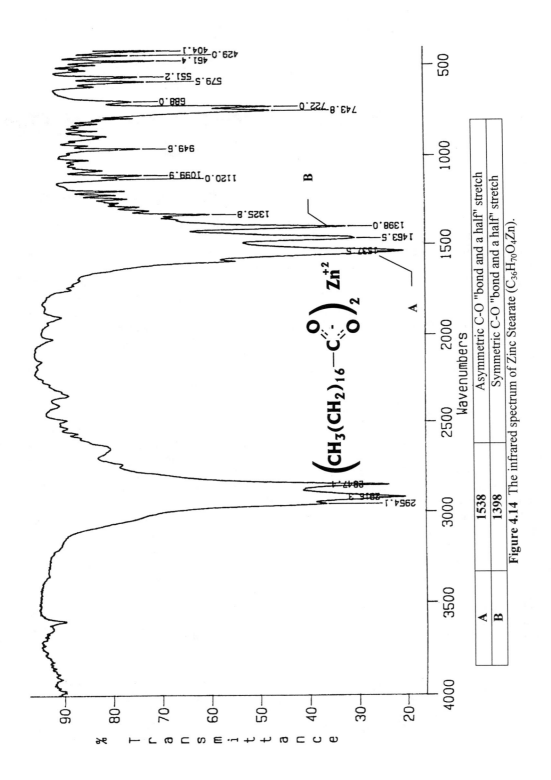

Figure 4.14 The infrared spectrum of Zinc Stearate ($C_{36}H_{70}O_4Zn$).

A	1538	Asymmetric C-O "bond and a half" stretch
B	1398	Symmetric C-O "bond and a half" stretch

C. Acid Anhydrides
1. Structure and Nomenclature

An *acid anhydride* is formed when two molecules of a carboxylic acid react to give an anhydride and a water molecule. This reaction, along with the structure of an acid anhydride, is shown in Figure 4.15.

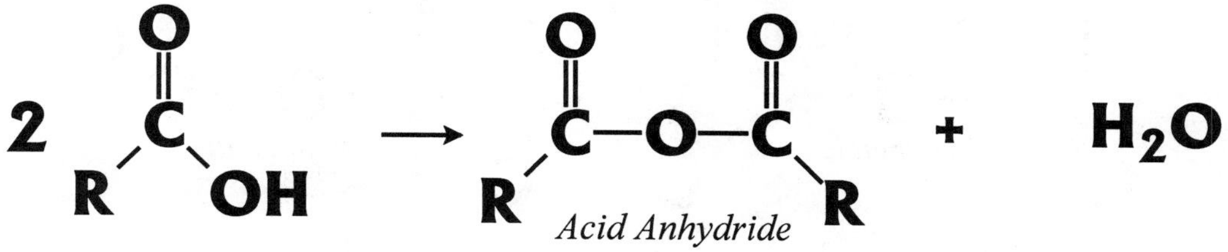

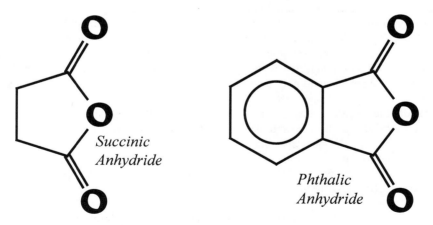

Figure 4.15 The reaction of two carboxylic acid molecules to form an acid anhydride. Note that the anhydride has two carbonyl groups.

By definition, an anhydride contains two carbonyl groups separated by an oxygen, and one carbon atom is attached to each carbonyl carbon. Anhydrides are named after the acid or acids from which they are made. For example, two molecules of acetic acid react to form acetic anhydride.

For many functional groups, the most commonly found saturated substituents are straight chain hydrocarbons, such as the propyl or hexyl groups. Cyclic hydrocarbon groups are not as commonly found as straight hydrocarbon chains. Anhydrides are unusual in that the most commonly found and economically important ones are cyclic. Examples are succinic anhydride and phthalic anhydride, whose structures are seen in Figure 4.16. Anhydrides such as these are common chemical intermediates and are used as crosslinking agents in polymers.

Figure 4.16 The structures of succinic and phthalic anhydrides.

Note that the carbonyl groups in phthalic anhydride are conjugated with the benzene ring. Whether an anhydride is cyclic or straight chain, whether it is conjugated or unconjugated, can be determined from the infrared spectrum of these materials.

2. Infrared Spectroscopy of Acid Anhydrides

The unique structural feature of an anhydride is the presence of two carbonyl groups linked to each other by an oxygen. This structural feature gives rise to two carbonyl stretching bands.

106 The Carbonyl Functional Group

Additionally, the C-O bond in an anhydride gives rise to a C-O stretch. The two infrared bands of an anhydride due to C=O stretching involve the two carbonyl groups stretching symmetrically (in phase with each other) and asymmetrically (out of phase with each other). Examples of these vibrations are seen in Figure 4.17.

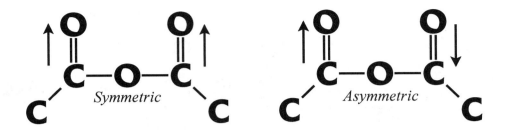

Figure 4.17 The symmetric and asymmetric carbonyl stretching vibrations of the anhydride functional group.

For most of the functional groups examined so far in this book, asymmetric stretches have occurred at higher wavenumber and had higher intensity than symmetric stretches. Anhydrides are an exception; the symmetric stretch occurs at higher wavenumber than the asymmetric stretch. For straight chain saturated anhydrides, the symmetric C=O stretch is found at 1820±5 cm^{-1} and the asymmetric stretch is found at 1750 ±5 cm^{-1}. These bands are seen in Figure 4.18, which is the spectrum of valeric anhydride, at 1819 and 1752 cm^{-1}, respectively. If a straight chain anhydride is conjugated, then the carbonyl stretches fall at 1775±5 cm^{-1} and 1720±5 cm^{-1}, respectively. Note in Figure 4.18 that the higher wavenumber symmetric C=O stretch is more intense than the lower wavenumber asymmetric stretch. This is typical of noncyclic anhydrides, and the intensity ratio of these two bands can be used to determine whether an anhydride is cyclic.

Cyclic anhydrides such as the ones seen in Figure 4.16 have their own unique C=O stretching vibrations. For a saturated cyclic anhydride, such as succinic anhydride, the symmetric and asymmetric C=O stretches are found from 1870 to 1845 cm^{-1} and from 1800 to 1775 cm^{-1}. For a conjugated cyclic anhydride, such as phthalic anhydride, the symmetric and asymmetric carbonyl stretch ranges are 1860 to 1840 cm^{-1} and 1770±10 cm^{-1}. The distinguishing characteristic of a cyclic anhydride is that the lower wavenumber asymmetric stretch is more intense than the higher wavenumber symmetric stretch (the opposite is true for straight chain anhydrides).

Other bands due to anhydrides are from the stretching vibrations of the C-O-C and C-C-O groups. These bands can be very intense. For a straight chain anhydride, one of these bands appears from 1060 to 1035 cm^{-1}. In the spectrum of valeric anhydride, the C-O stretch is the most intense band in the spectrum at 1038 cm^{-1}. For cyclic anhydrides, there are two bands, at 960-880 and 1300-1000 cm^{-1}. Tables 4.5 and 4.6 summarize the group wavenumbers for noncyclic and cyclic anhydrides.

Table 4.5 The Group Wavenumbers for Noncyclic Acid Anhydrides (all numbers in cm^{-1})

Vibration	Wavenumber Range
Saturated Symmetric C=O stretch	1820±5 (stronger)
Saturated Asymmetric C=O stretch	1750±5 (weaker)
Unsaturated Symmetric C=O stretch	1775±5 (stronger)
Unsaturated Asymmetric C=O stretch	1720±5 (weaker)
C-O Stretch	1060-1035

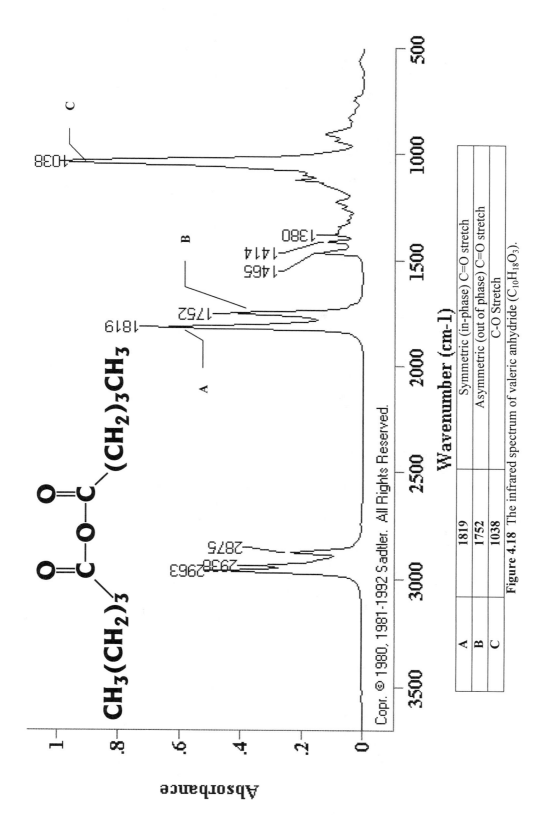

Figure 4.18 The infrared spectrum of valeric anhydride ($C_{10}H_{18}O_3$).

A	1819	Symmetric (in-phase) C=O stretch
B	1752	Asymmetric (out of phase) C=O stretch
C	1038	C-O Stretch

107

108 The Carbonyl Functional Group

Table 4.6 The Group Wavenumbers for Cyclic Acid Anhydrides (all numbers in cm^{-1})

Vibration	Wavenumber Range
Saturated Symmetric C=O stretch	1870-1845 (weaker)
Saturated Asymmetric C=O stretch	1800-1775 (stronger)
Unsaturated Symmetric C=O stretch	1860-1840 (weaker)
Unsaturated Asymmetric C=O stretch	1760-1780 (stronger)
C-O and C-C Stretches	960-880, 1300-1000

V. Esters

A. Structure and Nomenclature

Esters can also be thought of as derivatives of carboxylic acids, since they can be made by the reaction of a carboxylic acid and an alcohol. Esters are an important functional group; many of the flavoring agents in our food are esters. Biological compounds such as fats contain ester linkages, and billions of pounds of polyester are manufactured every year. Esters have a series of strong infrared bands which makes them easy to detect by infrared spectroscopy. The structural framework of an ester is shown in Figure 4.19.

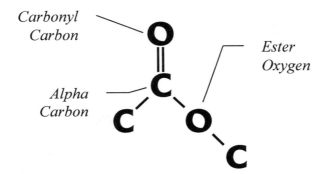

Figure 4.19 The skeletal framework of the ester functional group.

Note that the carbonyl carbon in an ester has a doubly bound and singly bound oxygen attached to it. The oxygen involved in the C-O single bond in an ester is called the *ester oxygen*, and has two carbon atoms singly bound to it. The carbon atom directly attached to the carbonyl carbon in an ester is called the *alpha carbon*. If the alpha carbon is saturated, then we have a *saturated ester*. If the alpha carbon is part of a benzene ring, then we have an *aromatic ester*. Note that it is the alpha carbon that determines if an ester is saturated or aromatic. The carbons attached to the ester oxygen play no role in determining the type of ester. Like all carbonyl containing functional groups, conjugation takes place when a benzene ring is directly attached to the carbonyl carbon of an ester.

B. The Infrared Spectra of Esters: The Rule of Three

All esters give rise to three strong infrared bands that appear at approximately 1700, 1200, and 1100 cm^{-1}. This pattern of bands is diagnostic for esters, and is called the "rule of three." The first of the rule of three bands is due to the C=O stretch of the ester group, and is similar to the other C=O stretches discussed in this chapter. The second of the rule of three bands, the one that appears near 1200 cm^{-1}, is due to the asymmetric stretching of the C-C and C-O bonds attached to the carbonyl carbon. This vibration is illustrated in Figure 4.20 and is called the "C-C-O" stretch. It is similar to the C-C-C stretch in ketones. Note that the C-C-O vibration involves the left side of the ester functional group.

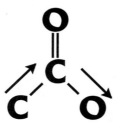

Figure 4.20 The C-C-O vibration of the ester group.

The third of these bands, which appears around 1100 cm^{-1}, is due to a vibration involving the ester oxygen and the next two carbons in the hydrocarbon chain attached to it. In this vibration, the O-C and C-O bonds stretch asymmetrically. This vibration is called the "O-C-C stretch" and involves the right-hand side of the functional group, as shown in Figure 4.21. This vibration is very similar to the C-O stretch of alcohols discussed in Chapter 3.

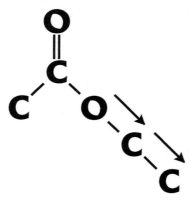

Figure 4.21 The O-C-C vibration of the ester functional group.

Saturated and aromatic esters both follow the rule of three. However, each of these types of ester has a unique set of infrared bands by which they can be identified. The next two sections will cover the spectroscopy of each type of ester in detail.

C. Saturated Esters

The spectrum of a saturated ester, ethyl acetate, is shown in Figure 4.22. Ethyl acetate is made by the reaction of acetic acid and ethyl alcohol, is used as an artificial fruit flavoring, and is found naturally in apples and pears. Acetate esters are made from acetic acid and contain the CH_3-C=O moiety.

Using the rule of three, look at the ester spectrum in Figure 4.22 and find the bands due to the ester. The band at 1742 cm^{-1} is the carbonyl stretch, which is typically found for saturated esters from 1750 to 1735 cm^{-1}. This carbonyl stretching region is a little higher than carbonyl stretches due to ketones, aldehydes, and carboxylic acids. As such, any carbonyl stretch greater than 1735 cm^{-1} should first be considered as being from a saturated ester.

The band at 1241 cm^{-1} in the spectrum of ethyl acetate is due to the C-C-O stretch. The position of this C-C-O stretch is unique to acetates, and falls at ~1240 cm^{-1}. All other saturated esters exhibit their C-C-O stretch from 1210 to 1160 cm^{-1}. This band is very strong, sometimes stronger than the ester C=O stretching band. The band at 1047 cm^{-1} in Figure 4.22 is the O-C-C stretch. For all saturated esters, this vibration appears between 1100 and 1030 cm^{-1}. A summary of the rule of 3 bands for a saturated ester is given in Table 4.7.

110 The Carbonyl Functional Group

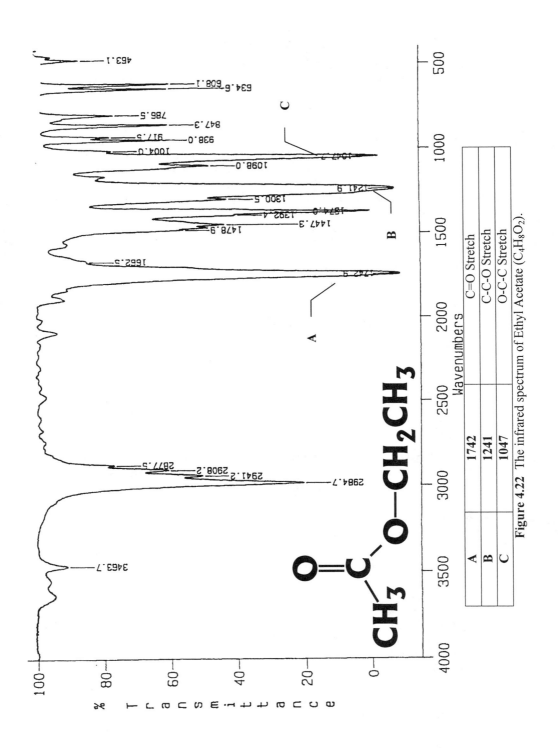

Figure 4.22 The infrared spectrum of Ethyl Acetate ($C_4H_8O_2$).

A	1742	C=O Stretch
B	1241	C-C-O Stretch
C	1047	O-C-C Stretch

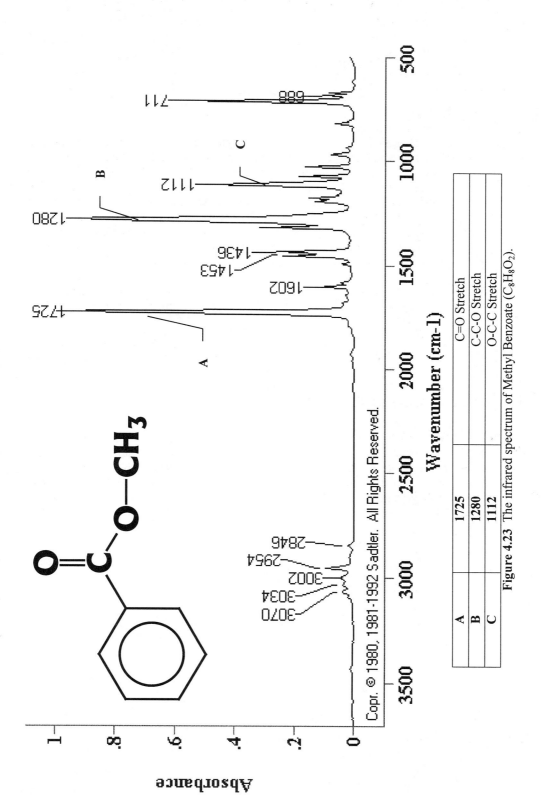

Figure 4.23 The infrared spectrum of Methyl Benzoate ($C_8H_8O_2$).

A	1725	C=O Stretch
B	1280	C-C-O Stretch
C	1112	O-C-C Stretch

Table 4.7 **The Rule of Three Bands For Saturated Esters** (all numbers in cm^{-1})

Vibration	Wavenumber Range
C=O Stretch	1750-1735
C-C-O Stretch	1210-1160
Acetate C-C-O Stretch	~1240
O-C-C Stretch	1100-1030

D. Aromatic Esters

Conjugation lowers the wavenumber of the ester C=O stretch by about 30 cm^{-1}. This is illustrated in Figure 4.23, which shows the infrared spectrum of methyl benzoate. This material is made from benzoic acid and methanol. The C=O stretch of this molecule falls at 1725 cm^{-1}, and the general range for the C=O stretch of aromatic esters is 1730 to 1715 cm^{-1}. The C-C-O stretch of methyl benzoate falls at 1280 cm^{-1}, and this vibration in aromatic esters is generally seen from 1310 to 1250 cm^{-1}. Finally, the O-C-C stretch of aromatic esters falls between 1130 and 1100 cm^{-1}, and is seen in Figure 4.23 at 1112 cm^{-1}. Note that the C=O, C-C-O, and O-C-C bands for aromatic esters fall into different ranges than for saturated esters. Therefore, it is easy to distinguish between aromatic and saturated esters based on the positions of the rule of three bands. Table 4.8 summarizes the group wavenumbers for aromatic esters.

Table 4.8 **The Rule of Three Bands For Aromatic Esters** (all numbers in cm^{-1})

Vibration	Wavenumber Range
C=O Stretch	1730-1715
C-C-O Stretch	1310-1250
O-C-C Stretch	1130-1000

VI. Organic Carbonates

A. Structure and Nomenclature

The carbonate functional group has the chemical formula CO_3. In inorganic carbonates the CO_3 group has a charge of minus two and is coordinated with two positively charged ions, normally metal ions. The spectroscopy of inorganic carbonates is discussed in Chapter 7. Organic carbonates have a totally covalent bonding scheme and have the structural framework seen in Figure 4.24.

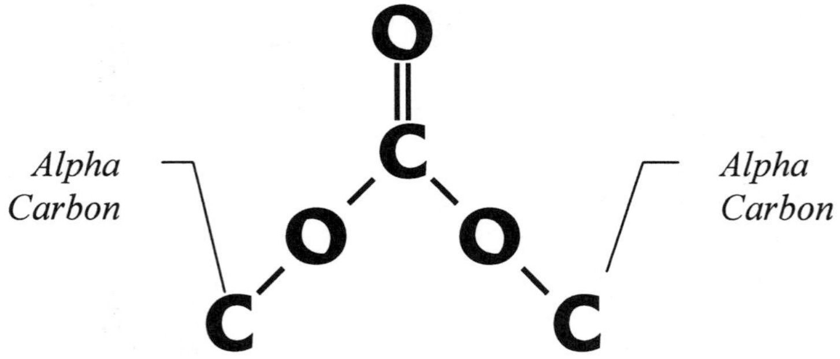

Figure 4.24 The structural framework of the organic carbonate functional group.

Note that a carbonate contains a C=O double bond, but has two C-O single bonds. The carbon atoms singly bonded to the oxygens in a carbonate are called *alpha carbons*.

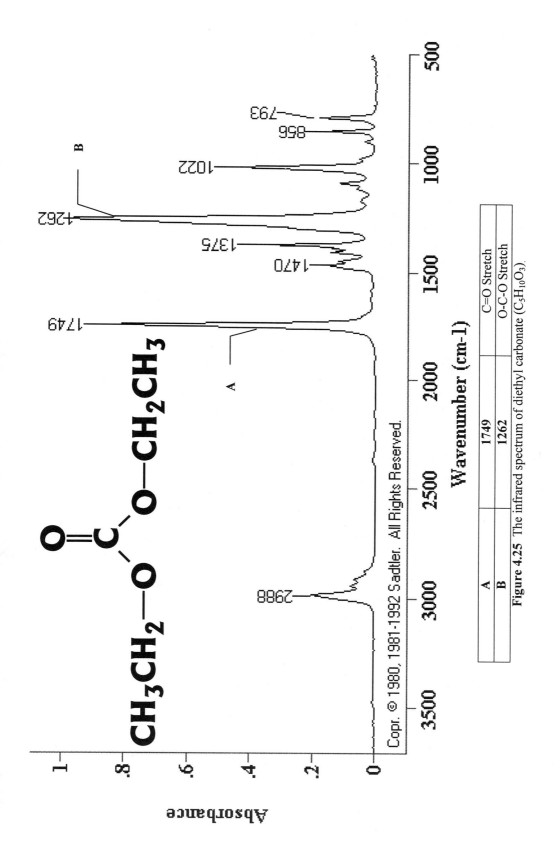

Figure 4.25 The infrared spectrum of diethyl carbonate ($C_5H_{10}O_3$).

A	1749	C=O Stretch
B	1262	O-C-O Stretch

In a *saturated carbonate*, both alpha carbons are saturated and are sometimes called dialkyl carbonates. An *aromatic carbonate* has alpha carbons that are both members of benzene rings. These are sometimes called diaryl carbonates. In a *mixed carbonate,* one alpha carbon is saturated and one alpha carbon is aromatic. Mixed carbonates can also be called aryl/alkyl carbonates. Lexan, a polycarbonate, is an important engineering plastic. Millions of pounds of it are manufactured each year. Its spectrum is discussed in Chapter 8.

B. The Infrared Spectra of Organic Carbonates

The C=O stretch of saturated carbonates falls at 1740 ± 10 cm^{-1}. This is illustrated in Figure 4.25, where the C=O stretch of diethyl carbonate is seen at 1748 cm^{-1}. For a mixed carbonate, the C=O stretch occurs between 1790 and 1760 cm^{-1}. For aromatic carbonates, this vibration appears from 1820 to 1775 cm^{-1}. There is some overlap in the ranges of these different vibrations, but the ranges are sufficiently different that often the type of carbonate present can be determined from the position of the C=O stretch.

The second important group wavenumber for organic carbonates is the asymmetric stretching of the two C-O bonds. This vibration is called the "O-C-O" stretch, and is very similar to the C-C-C stretch of ketones and the C-C-O stretch of esters. Diethyl carbonate exhibits an intense O-C-O stretch at 1260 cm^{-1} and, for saturated carbonates, this band is found from 1280 to 1240 cm^{-1}. Mixed carbonates have O-C-O stretches at 1250 to 1210 cm^{-1}, and aromatic carbonates have a band from 1220 to 1205 cm^{-1}. Table 4.9 summarizes the C=O and O-C-O stretches of the different organic carbonates.

Table 4.9 Group Wavenumbers for Organic Carbonates (all numbers in cm^{-1})

Vibration	Saturated Carbonate	Mixed Carbonate	Aromatic Carbonate
C=O Stretch	1740±10	1790-1760	1820-1775
O-C-O Stretch	1280-1240	1250-1210	1220-1205

VII. Summary

All of the functional groups discussed in this chapter have the C=O linkage in common, and have a C=O stretching band that appears prominently in their infrared spectra. However, the position of the carbonyl stretch is usually not enough to determine the substituents on the carbonyl carbon. Therefore, it is necessary to look elsewhere in the spectrum to be able to distinguish between carbonyl containing functional groups. The following list summarizes this information.

1. Ketones have a carbonyl stretch around 1700 cm^{-1} and a C-C-C stretch around 1200 cm^{-1}. They have no other strong infrared bands and their spectra are relatively simple.

2. The distinguishing structural feature of an aldehyde is the aldehydic hydrogen, which is isolated from the rest of the molecule by the carbonyl group. The aldehydic hydrogen gives rise to a unique, low wavenumber C-H stretch between 2850 and 2700. Aldehydes also have a C=O stretch just above 1700 cm^{-1}. The combination of a low wavenumber C-H stretch and a carbonyl stretch are strong indicators of an aldehyde.

3. Carboxylic acids have an O-H bonded to the carbonyl carbon and, due to strong hydrogen bonding, have unusually wide bands. The O-H stretch of a carboxylic acid is hundreds of wavenumbers wide, and identifies a molecule as a carboxylic acid. However, strong, wide bands due to O-H bending, C=O and C-O stretching are also found in the spectrum of carboxylic acids. An O-H stretch combined with a C=O stretch are diagnostic for carboxylic acids.

4. Carboxylates and anhydrides are derivatives of carboxylic acids. Carboxylates have two "bond and a half" C-O linkages that give rise to the asymmetric and symmetric CO_2 stretches at 1650-1540 and 1450-1360 cm^{-1}. These two bands are strong and easy to see. Anhydrides contain two carbonyl groups, so have two C=O stretches. The position and relative intensity of these bands can tell whether an anhydride is straight chain or cyclic, and whether it is saturated or aromatic.

5. Esters follow the rule of three: a pattern of intense bands at ~1700, ~1200, and ~1100 cm^{-1}. These bands are due to C=O, C-C-O, and O-C-C stretching. Saturated and aromatic esters have different sets of these bands.

6. Organic carbonates contain a CO_3 moiety. The C=O and O-C-O stretching vibrations help distinguish between saturated, mixed, and aromatic carbonates.

The C=O linkage is one of the most common in organic chemistry, and one of the most easily detected by infrared spectroscopy. I hope that this chapter has helped to shed some light on the unique and interesting spectroscopy of these molecules.

Bibliography

L.J. Bellamy, *The Infrared Spectra of Complex Molecules, Third Edition*, Wiley, New York, 1975.

R. Silverstein, G. Bassler, T. Morrill, *Spectrometric Identification of Organic Compounds, Fourth Edition*, Wiley, New York, 1981.

D. Lin-Vien, N. Colthup, W. Fately, J. Grasselli, *Infrared and Raman Characteristic Frequencies of Organic Molecules*, Academic Press, Boston, 1991.

N. Colthup, L. Daly, S. Wiberley, *Introduction to Infrared and Raman Spectroscopy*, Academic Press, New York, 1990.

F. Miller, *Lecture Notes for Bowdoin College Infrared Short Course*, Brunswick, Maine, 1992.

B.C. Smith, *Fundamentals of Fourier Transform Infrared Spectroscopy*, CRC Press, Boca Raton, 1996.

H. Mantsch, D. Chapman, Eds., *Infrared Spectroscopy of Biomolecules*, Wiley, New York, 1996.

A. Streitweiser, C. Heathcock, *Introduction to Organic Chemistry*, Macmillan, New York, 1976.

S. Budavari, Editor, *The Merck Index, 11th Edition*, Merck & Co., Rahway, NJ, 1989.

R.C. Weast, Editor, *The CRC Handbook of Chemistry and Physics*, CRC Press, Boca Raton, 1987.

Problem Spectra

The following spectra are presented as an exercise for the reader. Using the techniques and knowledge obtained in this chapter, do your best to predict the complete molecular structure of each sample from its infrared spectrum. The exact positions of prominent peaks are contained in the table accompanying each spectrum. However, these peaks may or may not be useful in the final determination of a structure. The physical state of the sample is also stated (solid or liquid) along with the sampling method used.

All problem set spectra were obtained on modern day FTIR instruments at 4 or 8 cm^{-1} resolution and are of pure materials. It is much easier to learn the basics of interpretation by looking at the simple spectra of pure compounds than to tackle the much more complex problem of analyzing mixture spectra. Please realize that the determination of a complete molecular structure in each problem may not be possible. However, by analyzing each of these spectra methodically and patiently, you will be able to learn a lot from each of them. The correct structure and a discussion of the interpretation of each problem are included in an appendix at the end of the book. Good Luck!

116 The Carbonyl Functional Group

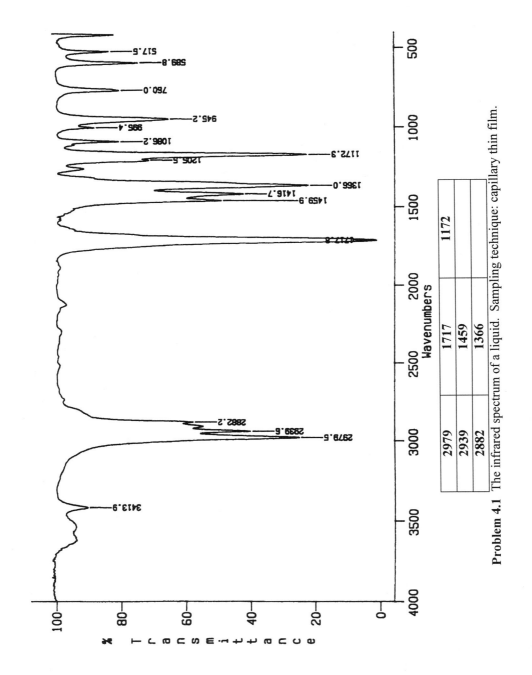

2979	1717	1172
2939	1459	
2882	1366	

Problem 4.1 The infrared spectrum of a liquid. Sampling technique: capillary thin film.

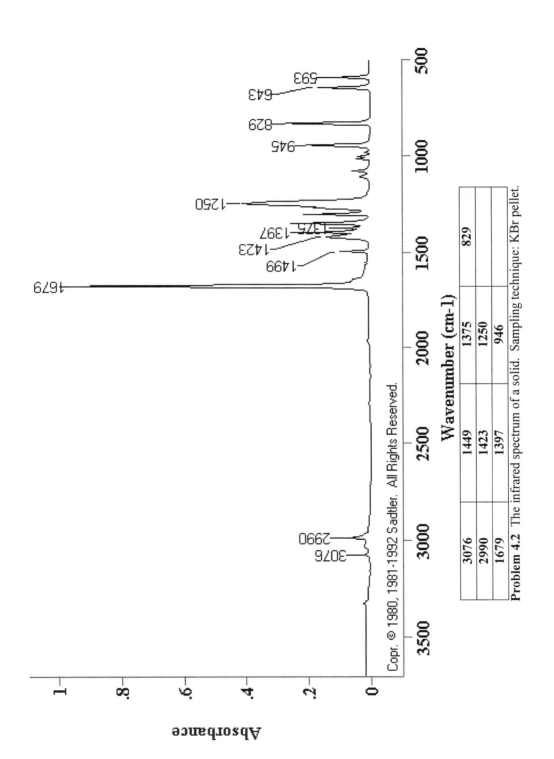

3076	1449	1375	829
2990	1423	1250	
1679	1397	946	

Problem 4.2 The infrared spectrum of a solid. Sampling technique: KBr pellet.

118   The Carbonyl Functional Group

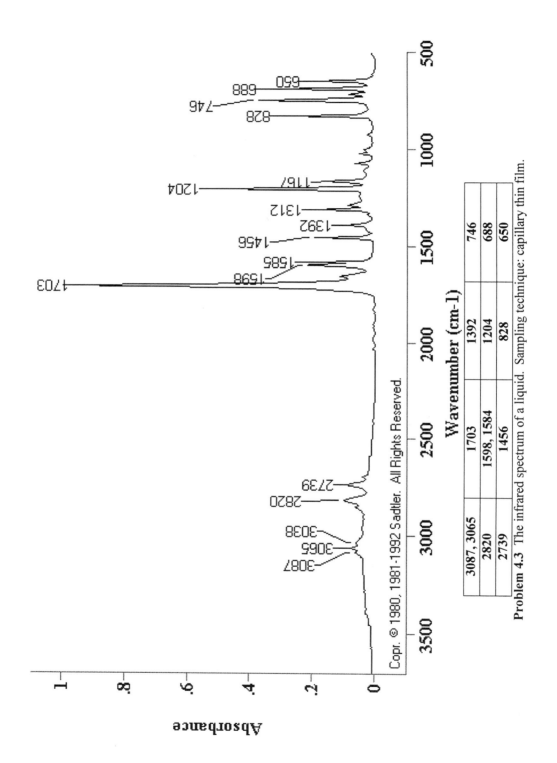

3087, 3065	1703	1392	746
2820	1598, 1584	1204	688
2739	1456	828	650

Problem 4.3 The infrared spectrum of a liquid. Sampling technique: capillary thin film.

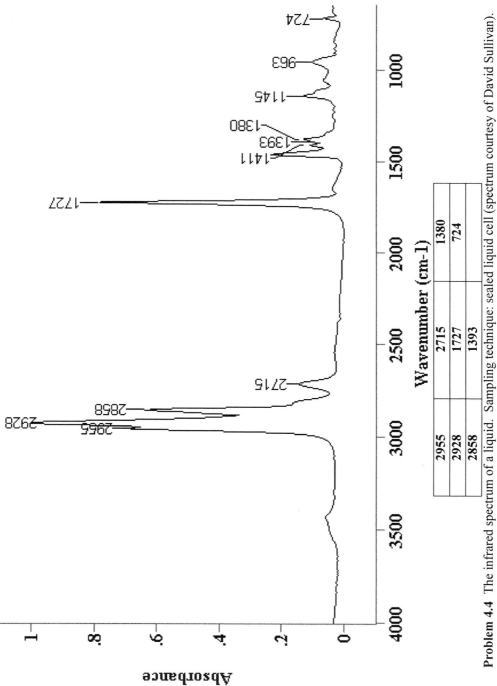

2955	2715	1380
2928	1727	724
2858	1393	

Problem 4.4 The infrared spectrum of a liquid. Sampling technique: sealed liquid cell (spectrum courtesy of David Sullivan).

120 The Carbonyl Functional Group

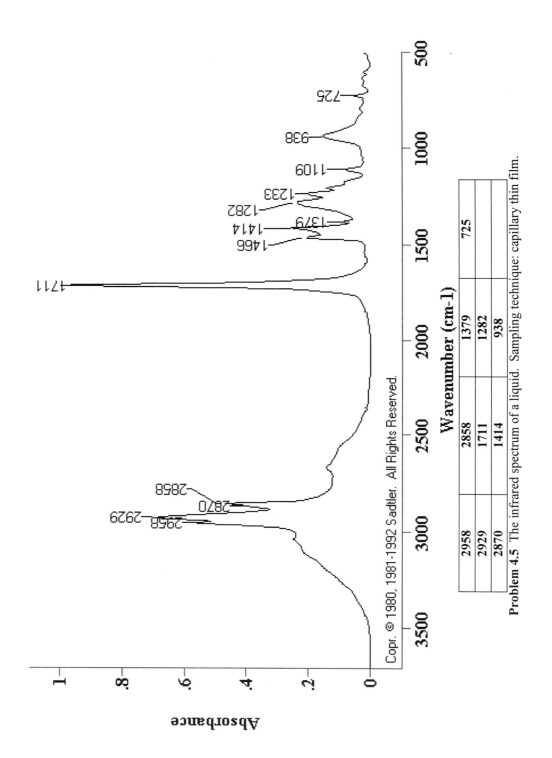

2958	2858	1379	725
2929	1711	1282	
2870	1414	938	

Problem 4.5 The infrared spectrum of a liquid. Sampling technique: capillary thin film.

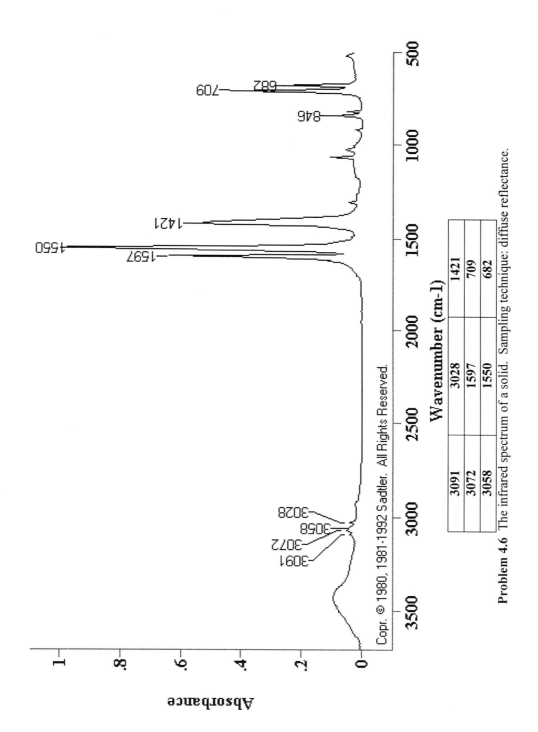

Problem 4.6 The infrared spectrum of a solid. Sampling technique: diffuse reflectance.

3091	3028	1421
3072	1597	709
3058	1550	682

122 The Carbonyl Functional Group

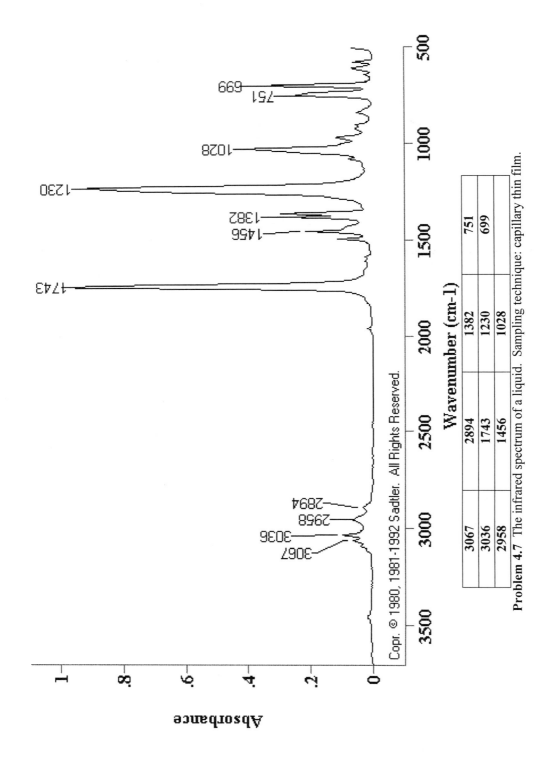

Problem 4.7 The infrared spectrum of a liquid. Sampling technique: capillary thin film.

3067	2894	1382	751
3036	1743	1230	699
2958	1456	1028	

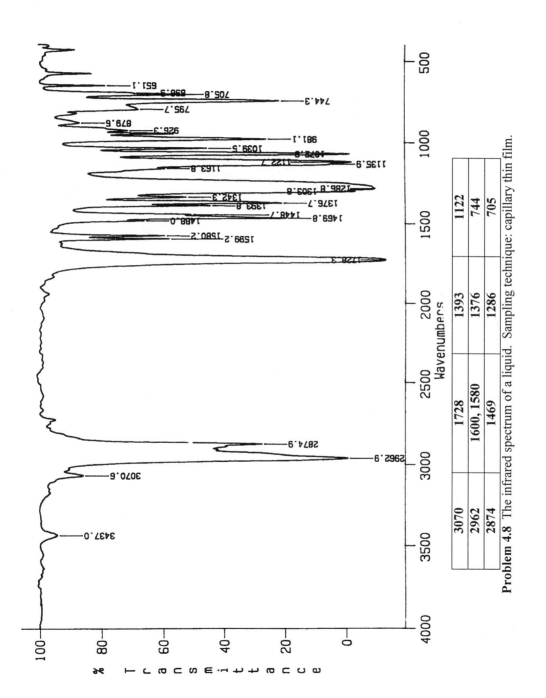

Problem 4.8 The infrared spectrum of a liquid. Sampling technique: capillary thin film.

3070	1728	1393	1122
2962	1600, 1580	1376	744
2874	1469	1286	705

124 The Carbonyl Functional Group

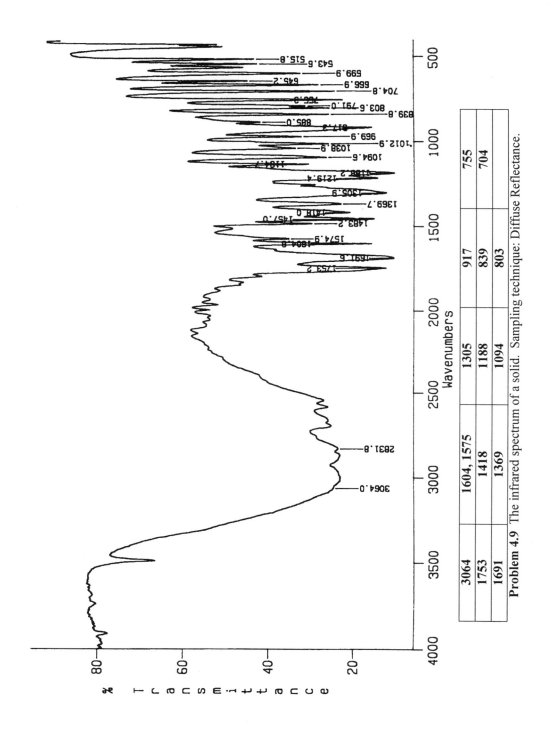

3064	1604, 1575	1305	917	755
1753	1418	1188	839	704
1691	1369	1094	803	

Problem 4.9 The infrared spectrum of a solid. Sampling technique: Diffuse Reflectance.

Chapter 5

Organic Nitrogen Compounds

I. Introduction

Up to this point, we have restricted our discussion to the infrared spectra of molecules that contain only carbon, hydrogen, and oxygen. However, there are a number of important organic functional groups that contain nitrogen. Any functional group that contains a carbonyl group with a nitrogen attached to the carbonyl carbon (amide, imides) has been included here rather than in the chapter on carbonyl compounds (Chapter 4).

Carbon and nitrogen have a complex chemistry and can form carbon/nitrogen single, double, and triple bonds, denoted C-N, C=N, and C≡N. The C-N bond is, in some ways, analogous to the C-O bond discussed in Chapter 3. However, nitrogen has a lower electronegativity than oxygen, and C-N bonds have a smaller dipole moment than C-O bonds. This means bands involving C-N bonds are generally weaker than bands involving C-O or C=O bonds. Single bond C-N stretches give rise to bands that show up from 1400 to 1000 cm^{-1} and are medium-to-weak in intensity. This makes them marginally useful as group wavenumbers.

The carbon nitrogen double bond, C=N, also gives rise to infrared features. However, the C=N bond is somewhat reactive, and compounds containing it are rare. We will not discuss the spectroscopy of the C=N bond in this chapter (see the volumes listed in the bibliography for more details on the less common nitrogen functional groups). The carbon nitrogen triple bond, C≡N, is stable and its stretching vibration gives rise to a very good group wavenumber that will be discussed below. The plethora of organic nitrogen functional groups means there is no one vibration or infrared feature that all these groups share. Therefore, we will limit our discussion in this chapter to nitrogen-containing functional groups that the average scientist is most likely to encounter, that have commercial importance, or have good group wavenumbers.

II. Amides

A. Structure and Nomenclature

An *amide* contains a carbonyl group with a carbon atom on one side and a nitrogen atom on the other side. The structural framework of an amide is shown in Figure 5.1. The carbon in the C=O bond is called a carbonyl carbon. The carbon attached to the carbonyl is called the alpha carbon. The nitrogen atom in an amide is called the *amide nitrogen*.

Amides are derivatives of carboxylic acids and are named by dropping the "ic" suffix from the name of the parent acid and attaching the suffix "amide." For example, the amide made from acetic acid is called acetamide (Figure 5.1 is the skeletal framework of acetamide). Both the carbonyl carbon and the amide nitrogen can have hydrogens or hydrocarbon substituents attached to them. In naming an amide, it is necessary to state what substituent is attached to which substituent. Substituents on the nitrogen are named by adding an "N-" to the beginning of the amide's name.

126   Organic Nitrogen Compounds

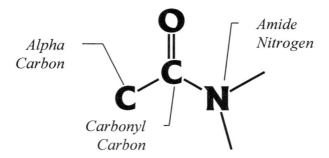

Figure 5.1 The structural framework of the amide functional group.

For example, N-methyl acetamide has a methyl group attached to the amide nitrogen, and N,N-dimethylacetamide has two methyl groups attached to the amide nitrogen. If there is no "N" at the beginning of the amide's name, the substituents are on the carbonyl carbon and are determined from the root of the amide's name. For example, the root name "prop-" always refers to a hydrocarbon chain with three carbon atoms. The last carbon of the chain is the carbonyl carbon. Therefore, a propionamide would have a CH_3CH_2 hydrocarbon chain attached to the carbonyl carbon.

All amides have at least one C-N bond, and are distinguished from each other by the total number of C-N bonds in the functional group. The skeletal frameworks of the three different types of amide are shown in Figure 5.2.

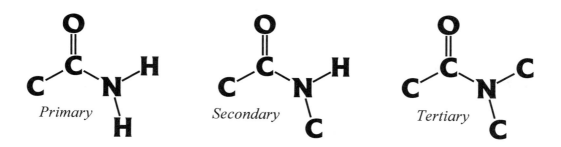

Figure 5.2 The skeletal frameworks of primary, secondary, and tertiary amides.

A *primary amide* has one C-N bond and two N-H bonds. A *secondary amide* has two C-N bonds and one N-H bond. A *tertiary amide* has three C-N bonds and no N-H bonds. Of the three varieties of amide, secondary amides are probably the most important. The backbone of every protein molecule in the world consists of a series of secondary amide linkages. Additionally, polymers made out of amides, such as nylon, contain secondary amide linkages. It is possible to distinguish between primary, secondary, and tertiary amides from their infrared spectra.

The nitrogen in an amide contains a pair of nonbonding electrons that are in a p-type orbital, which points above the plane of the amide. Recall from Chapter 4 that the carbonyl group has electrons in a similar orbital. These p-type electrons allow carbonyl groups to conjugate with benzene rings. The p-type electrons of the carbonyl group and the nitrogen atom in an amide can conjugate, allowing the electrons to "smear out" along the C-N bond. The spectroscopic effect of conjugation is to lower the C=O stretching wavenumber for amides compared to functional groups, such as ketones and esters. The chemical effect of conjugation is that the C-N bond is "stiffer" than expected and has some double bond character. This stiff C-N bond is strong enough to prevent the nitrogen from rotating about the C-N bond axis. As such, amides are "frozen" with one substituent closer to the carbonyl group than the other. In secondary amides, where the substituents are one hydrogen and one carbon atom, there exist cis and trans isomers of the molecule similar to the cis and trans isomers of the alkene functional group (see Chapter 2). The structure of the cis and trans isomers of secondary amides is shown in Figure 5.3.

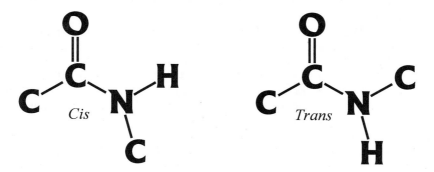

Figure 5.3 The skeletal frameworks of cis and trans secondary amides.

These two isomers have different infrared spectra, and both can be observed under certain conditions. However, the trans isomer is energetically more favorable and makes up >90% of the molecules in a sample of a secondary amide at room temperature. The secondary amide infrared bands discussed in this chapter will be only for the trans isomer.

The N-H bond in primary and secondary amides is similar to the O-H bond found in other functional groups. Nitrogen is more electronegative than hydrogen, and the electrons in an N-H bond have the highest probability of being found on the nitrogen atom. Consequently, there is a partial negative charge on the nitrogen and a partial positive charge on the hydrogen. In amides, the C=O group is polar, with a partial negative charge on the oxygen and a partial positive charge on the carbonyl carbon. The positive end of an N-H bond can coordinate with the negative end of a C=O bond to form an amide hydrogen bond. This is illustrated in Figure 5.4.

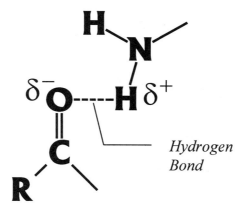

Figure 5.4 An example of hydrogen bonding in amides. Note that the N-H of one molecule bonds to the C=O of another molecule. The δ symbol indicates a partial charge.

The dipole moment of an N-H bond is smaller than that of an O-H bond, which means the hydrogen bonding in amides is weaker than in alcohols and carboxylic acids. The N-H stretching bands of an amide are of medium width and intensity compared to O-H stretches.

This fact can be used to distinguish between O-H and N-H stretches since they often fall in the same wavenumber range. The infrared bands of an amide are sensitive to the amount of hydrogen bonding taking place in a sample. For example, the positions of N-H stretch bands can shift over 100 cm^{-1} in going from pure amide to dilute solution. The infrared bands given for amides in this section will be only for the hydrogen-bonded form.

B. The Infrared Spectra of Primary, Secondary, and Tertiary Amides
1. Primary Amides

The number of N-H bonds in an amide equals the number of N-H stretching bands observed in its spectrum. Since primary amides have two N-H bonds, their spectra have two N-H stretching bands. These are seen in the spectrum of a primary amide, benzamide, shown in Figure 5.5. The two N-H stretches are at 3366 and 3170 cm^{-1}. They are of medium intensity and width. The wavenumber range for the N-H stretches of primary amides is 3370 to 3170 cm^{-1}. There must be two N-H stretches in this region for a molecule to be a primary amide.

Like any of the carbonyl-containing functional groups in Chapter 4, the C=O bond of an amide gives rise to a strong C=O stretching band. Because of conjugation between the carbonyl group and the amide nitrogen, the position of this C=O stretch is lower than in molecules such as ketones and esters. The general range for the C=O stretch of amides of all types is 1680 to 1630 cm^{-1}. In the spectrum of benzamide in Figure 5.5, the C=O stretch is the most intense band in the spectrum at 1656 cm^{-1}.

The NH$_2$ group in a primary amide can bend in addition to stretching. The in-plane bending vibration involves the H-N-H bond angle getting bigger and smaller, like the opening and closing of a pair of scissors. This vibration is called the NH$_2$ scissors mode, and is found from 1650 to 1620 cm^{-1}. The NH$_2$ scissors of benzamide occurs at 1622 cm^{-1}. Note that the wavenumber range for the C=O stretch and the NH$_2$ scissors overlap. Occasionally, the NH$_2$ bend will fall into the same place as the C=O stretch, giving rise to one intense band, or the C=O infrared band will have a shoulder on it due to the NH$_2$ scissors. The NH$_2$ part of a primary amide can also bend out of the plane defined by the functional group. It is sometimes called the "NH$_2$ wag" because the motion is similar to the wagging of a dog's tail. This out-of-plane bending band is typically very broad and is found from 750 to 600 cm^{-1}. In Figure 5.5, the NH$_2$ out-of-plane bend is a broad envelope centered at 700 cm^{-1}, and sharper bands due to the benzene ring are superimposed upon it. Primary amides contain one C-N bond, and the stretching of this bond gives rise to a band from 1430 to 1390 cm^{-1}. It is found in the spectrum of benzamide at 1398 cm^{-1}.

In summary, the infrared bands unique to a primary amide are two N-H stretches, the NH$_2$ scissors, and the position of the C-N stretch. The group wavenumbers for primary amides are summarized in Table 5.1.

Table 5.1 The Group Wavenumbers of Primary Amides (all numbers in cm^{-1}).

Vibration	Wavenumber Range
NH$_2$ Stretches	3370-3170 (2 bands)
C=O Stretch	1680-1630
NH$_2$ Scissors	1650-1620
C-N Stretch	1430-1390
NH$_2$ Wag	750-600 (broad)

2. Secondary Amides

Secondary amides are probably the most common and most important type of amide. Proteins, nylons, and other polymers contain secondary amide linkages. By definition, a secondary amide contains two C-N bonds and one N-H bond. Consequently, only one N-H stretching band is observed in the infrared spectra of secondary amides. This band appears between 3370 and 3170 cm^{-1}. The spectrum of Nylon-6,6 in Figure 5.6 shows the secondary amide N-H stretch at 3301 cm^{-1}. The carbonyl stretch of secondary amides appears in the same range as other amides, from 1680 to 1630 cm^{-1}.

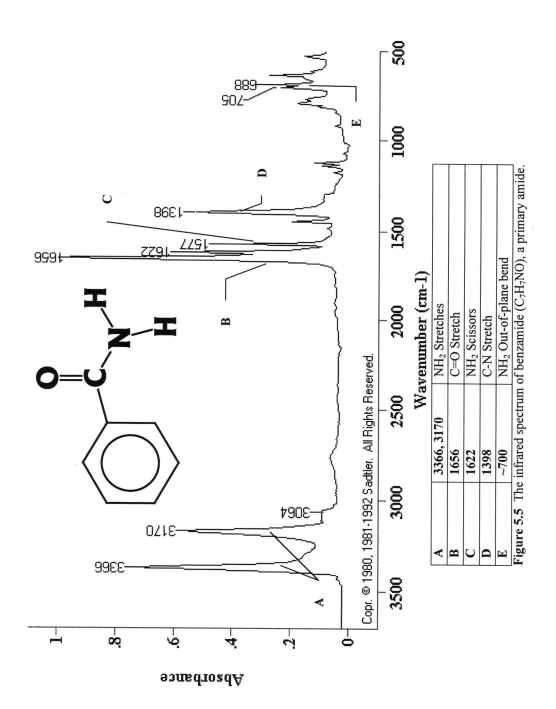

Figure 5.5 The infrared spectrum of benzamide (C_7H_7NO), a primary amide.

A	3366, 3170	NH_2 Stretches
B	1656	C=O Stretch
C	1622	NH_2 Scissors
D	1398	C-N Stretch
E	~700	NH_2 Out-of-plane bend

In Figure 5.6, the C=O stretch appears at 1641 cm^{-1}. This band is lower than in other carbonyl containing functional groups because of conjugation with the amide nitrogen.

The N-H moiety can bend as well as stretch and, like many other functional groups, there is an in-plane and out-of-plane bending vibration in secondary amides. The in-plane N-H bending vibration is an important group wavenumber. This vibration usually appears from 1570 to 1515 cm^{-1}. It is found in the spectrum of Nylon-6,6 at 1542 cm^{-1}.

The secondary amide in-plane N-H bend band is often intense, almost as intense as the C=O stretch itself. The only other functional groups that have strong bands around 1550 cm^{-1} are carboxylates (Chapter 4) and the nitro group (see below). Neither of these groups has a carbonyl stretch. Therefore, the combination of a carbonyl stretch and a strong N-H bending band between 1570 and 1515 cm^{-1} is diagnostic for secondary amides.

The in-plane N-H bend sometimes gives rise to an overtone band (see Chapter 1) at about twice the N-H bend fundamental at ~1550 cm^{-1}. This band appears just above 3000 cm^{-1}, in the region where unsaturated C-H stretches occur. The N-H bend overtone band is seen in the spectrum of Nylon-6,6 at 3081cm^{-1}. This band can be distinguished from C-H stretches because the N-H stretch is much broader than any C-H stretch.

The secondary amide out-of-plane N-H bend is not as important as the in-plane bend. It gives rise to a broad, medium-to-weak band from 750 to 680 cm^{-1}. This band appears in Figure 5.6 at 691 cm^{-1}. Note that the range for the primary and secondary amide out-of-plane N-H bends overlap, so these bands cannot be used to distinguish these types of amides.

The last group wavenumber for secondary amides is the C-N stretch. It appears in the spectrum of Nylon-6,6 at 1274 cm^{-1}. This band usually appears from 1310 to 1230 cm^{-1}. It appears much lower than the C-N stretch of primary amides because there is vibrational interaction (Fermi resonance, see Chapter 1) between the C-N stretch and N-H bend. Consequently, the N-H bend appears higher and the C-N stretch appears lower than expected.

To summarize, the important secondary amide infrared bands are a single N-H stretch, and carbonyl stretch combined with an intense N-H bend near 1550 cm^{-1}. Table 5.2 lists the important group wavenumbers for the secondary amide functional group.

Table 5.2 The Group Wavenumbers of Secondary Amides (all numbers in cm^{-1}).

Vibration	Wavenumber Range
N-H Stretch	3370-3170 (1 band)
C=O Stretch	1680-1630
N-H In-plane bend	1570-1515 (strong)
C-N Stretch	1310-1230
N-H out-of-plane bend	750-680 (broad)

3. Tertiary Amides

A tertiary amide has three C-N bonds and no N-H bonds. Consequently, there are no N-H stretching or bending bands in the spectra of tertiary amides. The only useful tertiary amide group wavenumber is the C=O stretch itself, which falls between 1680 and 1630 cm^{-1}, the same range as for all other types of amide. There is a group wavenumber for tertiary amides that are N,N-dimethyl substituted. This means that there are two methyl groups attached to the amide nitrogen. This group gives rise to a medium intensity band around 1505 cm^{-1} that is not observed in the spectra of other amides, and can be assigned as a C-N stretch. Table 5.3 summarizes the bands for tertiary amides.

Table 5.3 The Group Wavenumbers of Tertiary Amides (all numbers in cm^{-1}).

Vibration	Wavenumber Range
C=O Stretch	1680-1630
C-N(CH$_3$)$_2$ Stretch	~1505

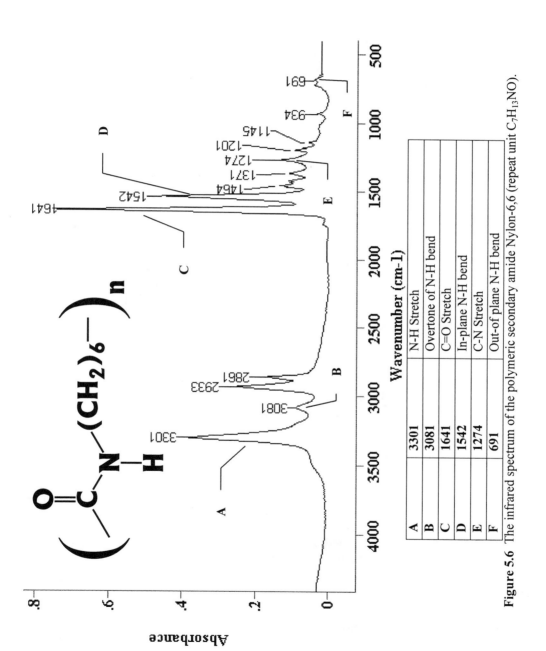

Figure 5.6 The infrared spectrum of the polymeric secondary amide Nylon-6,6 (repeat unit $C_7H_{13}NO$).

4. Proteins

Protein molecules are of great biological importance. Proteins are polymers whose repeat units are chosen from the 20 amino acids that exist in the human body. The sequence of amino acids in a protein is called its primary structure. Secondary amide bonds link the repeat units in a protein. Because of the stiffness of the amide C-N bond, protein chains can form three-dimensional structures such as helices, coils, and sheets. These comprise the secondary structure of a protein. The tertiary structure of a protein is determined by the positions in space of the side groups attached to the amino acid chain. The tertiary structure is involved in determining the chemistry of an enzyme, and subtle changes in it can destroy an enzyme's activity.

The infrared spectra can be used to identify protein secondary structure. For example, the C=O stretch of a helix is different from that of a sheet. The spectrum of a common protein, keratin, is shown in Figure 5.7. Keratin is the main component in hair, horn, claws, fingernails, and toe nails. The spectrum in Figure 5.7 is of wool fibers (sheep hair).

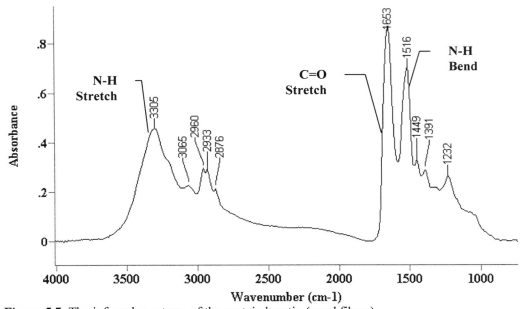

Figure 5.7 The infrared spectrum of the protein keratin (wool fibers).

Note the similarity between Figure 5.7 and the spectrum of Nylon-6,6 seen in Figure 5.6. Keratin has an N-H stretch at 3305 cm^{-1}, a carbonyl stretch at 1653 cm^{-1}, and an in-plane N-H bend at 1516 cm^{-1}. Additionally, the overtone of the N-H bend at 3065 cm^{-1} and the C-N stretch at 1232 cm^{-1} are visible. This spectrum has all the characteristics of a secondary amide. For more information on the infrared spectroscopy of proteins, consult the volumes on the infrared spectroscopy of biological molecules listed in the bibliography.

III. Imides

A. Structure and Nomenclature

Imides contain two carbonyl groups separated by a nitrogen atom. The skeletal framework of an imide is shown in Figure 5.8. The carbons in the C=O bond are carbonyl carbons, and the nitrogen atom separating the two carbonyl groups is called the *imide nitrogen*. Note that the two carbonyl carbons can have substituents that are the same or different, giving rise to symmetric and asymmetric imides. The substituent on the nitrogen atom can be a hydrogen atom or a carbon-bearing moiety. If an imide contains an N-H bond, its structure is very similar to that of a secondary amide.

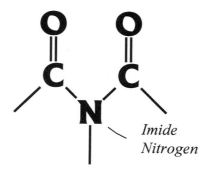

Figure 5.8 The skeletal framework of the imide functional group.

Like amides, imides can be made from carboxylic acids and are usually named after the acid(s) from which they are made. The "ic" suffix is dropped from the acid name and the suffix "imide" is added. For example, an imide made from phthalic acid is named phthalimide. If the nitrogen is substituted with something other than hydrogen, an "N-" and the name of the substituent is appended to the beginning of the name. For example, if there is a methyl group attached to the nitrogen in phthalimide, it is called N-methylphthalimide. Like acid anhydrides, the most important imides are the cyclic imides, such as phthalimide whose structure is seen along with its spectrum in Figure 5.9. Imides form an important family of engineering plastics called Kaptons®. The spectra of polyimides are discussed in Chapter 8.

B. The Spectroscopy of Imides

Straight chain imides have spectra very similar to secondary amides. If an imide has an N-H bond, its N-H stretch is found at 3200±50 cm^{-1}. The imide in-plane N-H bending band is found at ~1505 cm^{-1}. Straight chain imides give rise to one C=O stretching band, found from 1740 to 1670 cm^{-1}, and a C-N stretch from 1235 to 1165 cm^{-1}.

As mentioned above, cyclic imides are more common and more commercially important than straight chain imides. The spectrum of a cyclic imide, phthalimide, is seen in Figure 5.9. The N-H stretch in this spectrum is at 3205 cm^{-1}. There are two carbonyl stretches in the spectra of cyclic imides, corresponding to in-phase and out-of-phase stretching of the two C=O groups. These usually appear from 1790 to 1735 cm^{-1} and from 1750 to 1680 cm^{-1}. In the spectrum of phthalimide, the two carbonyl stretches appear at 1774 and 1745 cm^{-1}. Note that the band at lower wavenumber is more intense, which is typical of cyclic imides. The in-phase N-H bend and the C-N stretch in cyclic imides is weak to nonexistent. Straight chain imides have only one carbonyl stretch, while cyclic imides have two. This fact can be used to distinguish between these two types of imide. Although the carbonyl stretch of cyclic imides appears in the same wavenumber range as many other functional groups, it is unusual to see two carbonyl stretches between 1700 and 1800 cm^{-1}. The presence of two carbonyl stretches is a good diagnostic marker for cyclic imides. Table 5.4 summarizes the group wavenumbers for straight chain and cyclic imides.

Table 5.4 The Group Wavenumbers of Imides (all numbers in cm^{-1}).

Vibration	Straight Chain	Cyclic
N-H Stretch	3200±50	3200±50
C=O Stretch(es)	1740-1670 (1 band)	1790-1735, 1750-1680 (2 bands)
N-H In-plane bend	~1505	-
C-N Stretch	1235-1165	-

134   Organic Nitrogen Compounds

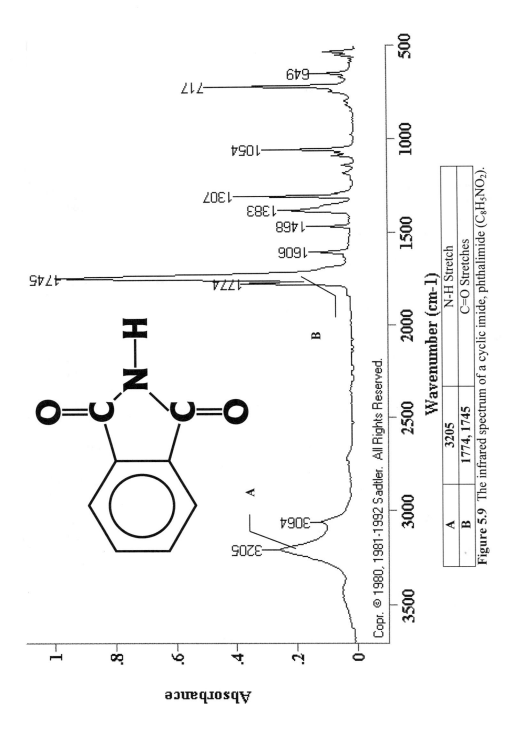

A	3205	N-H Stretch
B	1774, 1745	C=O Stretches

Figure 5.9 The infrared spectrum of a cyclic imide, phthalimide ($C_8H_5NO_2$).

IV. Amines

A. Structure and Nomenclature

Amines are one of the most commonly occurring organic nitrogen functional groups. Amines contain only carbon, hydrogen, and nitrogen, and have at least one C-N bond. The number of C-N bonds in an amine determines the type of amine. A molecule with one C-N bond is a *primary amine*, a molecule with two C-N bonds is a *secondary amine*, and a molecule with three C-N bonds is a *tertiary amine*. The substituents on the nitrogen other than carbon atoms are always hydrogen atoms. The number of N-H bonds in primary, secondary, and tertiary amines is 2, 1, and 0, respectively. Figure 5.10 shows the skeletal framework of the three different types of amine.

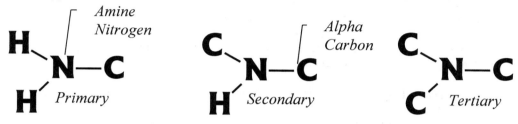

Figure 5.10 The skeletal frameworks of primary, secondary, and tertiary amines.

Any carbon attached to the nitrogen is called an alpha carbon, regardless of the type of amine. The nitrogen in an amine is called an *amine nitrogen*. Like many other functional groups, amines are divided into *saturated amines* and *aromatic amines* depending on the nature of the alpha carbons. Any time an aromatic ring is directly bonded to the nitrogen, it is considered an aromatic amine.

Amines are named by simply stating the name of the hydrocarbon groups attached to the nitrogen, followed by the word "amine." Thus, a primary amine with a methyl group attached to the nitrogen would be called "methylamine." If it is not clear whether a hydrocarbon moiety is attached to the nitrogen or another carbon, the prefix "N-" is used. Thus, a secondary amine with a methyl and propyl group present might be called "N-methylethylamine." This clearly states that the methyl group is directly attached to the nitrogen. Tertiary amines are often named with the prefix "N,N-" to denote which hydrocarbon moieties are directly attached to the nitrogen. Thus, a nitrogen with a propyl and two methyl groups attached to it would be named "N,N-dimethylpropylamine." Aromatic amines are often named after the simplest molecule of this type, *aniline*. Aniline consists of a benzene ring with an NH_2 group attached to it. Thus, a nitrogen atom with a benzene ring and an ethyl group attached to it would be called "N-ethylaniline." An aniline with two methyls attached to the nitrogen would be named "N,N-dimethylaniline."

Like the amide functional group discussed earlier in this chapter, the N-H bonds in primary and secondary amines are polar, with a partial positive charge on the hydrogen and a partial negative charge on the nitrogen. Because of these charges, the hydrogen of one amine molecule can coordinate with the nitrogen of a second amine molecule forming a hydrogen bond. An example of hydrogen bonding in amines is seen in Figure 5.11. Recall that, for alcohols, the impact upon the infrared spectrum of hydrogen bonding was to broaden those bands due to the O-H group. In a similar fashion, the infrared bands due to N-H bonds are broadened due to hydrogen bonding.

The stretching bands of N-H and O-H bonds often fall in the same region. However, the hydrogen bonding in amines is weaker than in alcohols, so N-H stretching bands are not as broad or intense as O-H stretching bands.

136   Organic Nitrogen Compounds

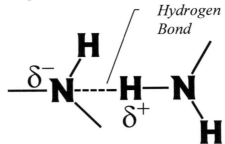

Figure 5.11 An example of hydrogen bonding in amines.

The positions of N-H stretching and bending bands are sensitive to the amount of hydrogen bonding in a sample. Thus, the positions of these bands are different for pure samples of an amine versus a dilute solution. However, most of the time we will encounter amines in the "associated" or hydrogen bonded state. Therefore, all the infrared band positions given for amines in this section will be for hydrogen-bonded molecules.

B. The Infrared Spectra of Primary, Secondary, and Tertiary Amines
1. Primary Amines
The unique structural feature of a primary amine is the presence of an NH_2 group. This moiety can stretch and bend giving rise to unique infrared absorbances. The three most important vibrations of the primary amine functional group are shown in Figure 5.12.

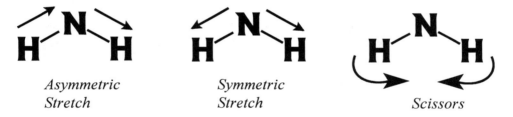

Figure 5.12 The asymmetric NH_2 stretch, symmetric NH_2 stretch, and NH_2 scissors mode of the primary amine functional group.

The asymmetric NH_2 stretch involves one hydrogen moving toward the nitrogen while the other hydrogen moves away from the nitrogen. The spectrum of a primary amine, propyl amine, is shown in Figure 5.13. The asymmetric NH_2 stretch in this spectrum appears at 3369 cm^{-1}. For saturated primary amines, this vibration generally appears from 3380 to 3350 cm^{-1}. The symmetric NH_2 stretch of a saturated primary amine involves both hydrogens simultaneously moving toward or away from the nitrogen. For a saturated primary amine, this vibration gives a band from 3310 to 3280 cm^{-1}. In the spectrum of propylamine, this band appears at 3298 cm^{-1}.

The N-H stretches of aromatic primary amines fall into different ranges than those of saturated primary amines. Specifically, the asymmetric NH_2 stretch appears from 3500 to 3420 cm^{-1} and the aromatic amine symmetric NH_2 stretch appears from 3420 to 3340 cm^{-1}. The positions of these N-H stretches can sometimes be used to determine whether or not there is a benzene ring attached to the amine nitrogen.

The O-H stretches of alcohols and carboxylic acids occur in the same wavenumber region as the NH_2 stretches of primary amines. Bands due to an amine will be weaker and narrower than O-H stretching bands. The stretching bands of the NH_2 group are best characterized as having medium width and medium intensity.

137

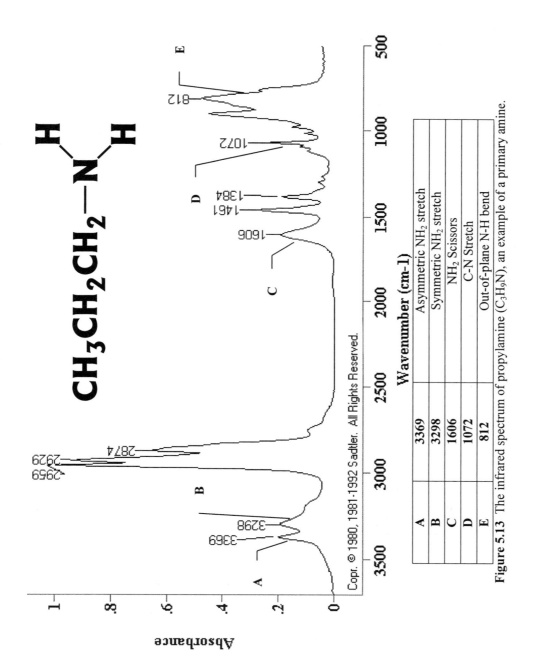

Figure 5.13 The infrared spectrum of propylamine (C_3H_9N), an example of a primary amine.

A	3369	Asymmetric NH_2 stretch
B	3298	Symmetric NH_2 stretch
C	1606	NH_2 Scissors
D	1072	C-N Stretch
E	812	Out-of-plane N-H bend

138   Organic Nitrogen Compounds

The bending of the primary amine NH$_2$ group can be in-plane or out-of-plane. The in-plane bending mode is a "scissoring" motion where the H-N-H bond angle gets bigger and smaller, like the opening and closing of a pair of scissors. This scissoring vibration is illustrated in Figure 5.12. For primary amines, this band falls from 1650 to 1580 cm^{-1}. It is seen in the spectrum of propylamine at 1606 cm^{-1}. Note that the NH$_2$ bending band in Figure 5.13 is broader than most bands that appear in this region, such as carbonyl stretches and aromatic ring modes.

The out-of-plane NH$_2$ bend involves both hydrogens "wagging" above and below the plane defined by the C-N bond. For primary amines, this band appears from 850 to 750 cm^{-1} and is seen in Figure 5.13 at 812 cm^{-1}. Note how broad this band is compared to other bands in the spectrum.

Since a primary amine has one C-N bond, it has an infrared band due to the C-N stretch. This band is seen in Figure 5.13 at 1072 cm^{-1}. The primary amine C-N stretching vibration for saturated molecules occurs from 1250 and 1020 cm^{-1}. For primary aromatic amines, the C-N stretch falls between 1350 and 1250 cm^{-1}. Note that the position of this band can be used to distinguish between saturated and aromatic amines.

The problem with the C-N stretching group wavenumber is that it falls in the same range as bands involving C-O bonds, but it is only medium to weak in intensity. Therefore, more intense C-O stretching bands can mask bands due to C-N stretching. Additionally, if the alpha carbon in a primary amine is branched, several extra bands due to C-C and C-C-N vibrations can appear between 1250 and 1020 cm^{-1}, making the assignment of the exact C-N stretching band somewhat confusing.

To summarize, primary amines have two NH$_2$ stretching bands at high wavenumber, an NH$_2$ scissoring band that is uniquely broad for its wavenumber range and a C-N stretch that can be used to distinguish saturated and aromatic amines. Table 5.5 summarizes the important group wavenumbers for primary amines.

Table 5.5 The Group Wavenumbers of the Primary Amine Group (all numbers in cm^{-1}).

Vibration	Saturated	Aromatic
NH$_2$ Asymmetric stretch	3380-3350	3500-3420
NH$_2$ Symmetric stretch	3310-3280	3420-3340
NH$_2$ Scissors	1650-1580	1650-1580
C-N Stretch	1250-1020	1350-1250
NH$_2$ Out-of-plane bend	850-750	850-750

2. Secondary Amines

By definition, the secondary amine functional group has two C-N bonds and one N-H bond Consequently, it exhibits only one N-H stretching band. For saturated secondary amines, the N-H stretch appears between 3320 and 3280 cm^{-1}. This band is at 3284 cm^{-1} in the spectrum of N-methylcyclohexylamine, seen in Figure 5.14. For secondary aromatic amines, the N-H stretch is found near 3400 cm^{-1}. Note that the N-H stretching band has medium intensity and medium width.

In theory, the N-H bond of a secondary amine should exhibit a band due to in-plane bending. However, this band is variable in its intensity and position, making it a poor group wavenumber. Fortunately, the out-of-plane N-H bend of a secondary amine, sometimes called the N-H "wag," gives rise to a good group wavenumber. The band due to this vibration appears between 750 and 700 cm^{-1} in the spectra of secondary amines. It appears in the spectrum in Figure 5.14 at 736 cm^{-1}. Note that the out-of-plane NH$_2$ bending vibration (the primary amine "wag") appears from 850 to 750 cm^{-1}, as compared to 750 to 700 cm^{-1} for a secondary amine wag. Thus, the position of this band can be used in addition to the number of N-H stretching vibrations to distinguish between primary and secondary amines.

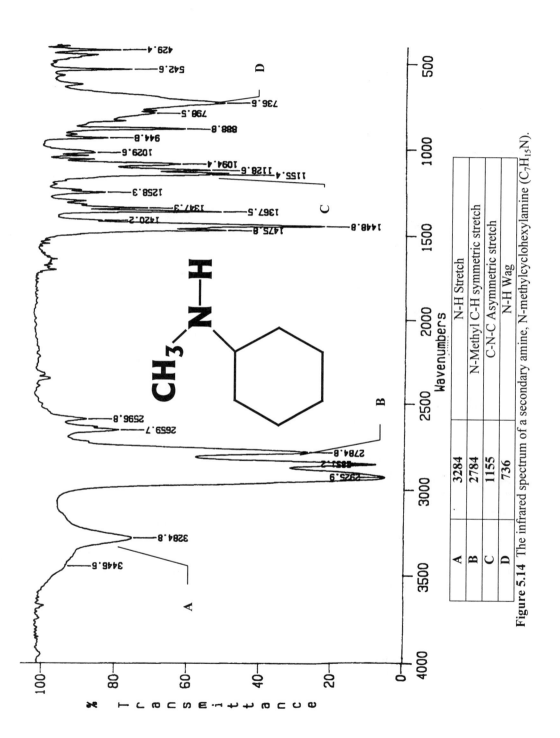

Figure 5.14 The infrared spectrum of a secondary amine, N-methylcyclohexylamine ($C_7H_{15}N$).

A	3284	N-H Stretch
B	2784	N-Methyl C-H symmetric stretch
C	1155	C-N-C Asymmetric stretch
D	736	N-H Wag

Since a secondary amine has two C-N bonds, a vibration involving the asymmetric stretching of these two bonds can occur, called the "C-N-C asymmetric stretch." This vibration is illustrated in Figure 5.15.

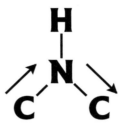

Figure 5.15 The C-N-C asymmetric stretching vibration of a secondary amine.

Note that one C-N bond stretches while the other contracts. The infrared band due to this vibration appears between 1180 and 1130 cm^{-1} for saturated secondary amines. It can be seen in the spectrum of N-methylcyclohexylamine at 1155 cm^{-1}. For aromatic secondary amines, the C-N-C asymmetric stretch is found between 1350 and 1250 cm^{-1}. Note that saturated and aromatic secondary amines can be distinguished by the position of this band.

If there is branching at the alpha carbons of a secondary amine, or if the two alpha carbons are of different varieties (primary, secondary, or tertiary), there may be a number of bands involving C-C and C-N vibrations present. This may make it difficult to identify which band is the C-N-C asymmetric stretch.

In summary, the presence of one N-H stretching band and an N-H wagging band from 750 to 700 cm^{-1} mark the infrared spectra of secondary amines. The group wavenumbers of secondary amines are listed in Table 5.6.

Table 5.6 The Group Wavenumbers of the Secondary Amine Group (all numbers in cm^{-1}).

Vibration	Saturated	Aromatic
N-H Stretch	3320-3280	~3400
C-N Stretch	1180-1130	1350-1250
N-H Wag	750-700	750-700

3. Tertiary Amines

Tertiary amines have three C-N bonds and no N-H bonds. Consequently, there are no N-H stretching or bending bands in the spectra of tertiary amines. The only infrared bands specific to tertiary amines are C-N stretches, which occur from 1250 to 1020 cm^{-1}. However, many other functional groups have bands in this range, which makes it very difficult to detect tertiary amines via infrared spectroscopy. Often the spectra of tertiary amines consist of the bands due to the hydrocarbon substituents, with a collection of nondescript medium intensity bands between 1200 and 100 cm^{-1}. Methyl groups directly attached to a nitrogen have a unique C-H stretching band, which can sometimes be used to signal the presence of the tertiary amine group (see next section).

The best way to determine whether or not a sample contains a tertiary amine is to make a derivative of the suspected amine, then take its spectrum. To make the derivative, mix 2 drops of the sample with 1 drop of a 50/50 mixture of HCl in alcohol. If present, the tertiary amine will react to form its hydrochloride salt and precipitate from solution. The precipitate has an unusual band at 2600 cm^{-1} that can be used to help confirm the presence of the tertiary amine in the original sample.

4. Methyl Groups Attached to an Amine Nitrogen

Methyl (CH_3) groups are commonly found attached to amine nitrogens. The C-H stretching force constants of these methyl groups are lower than methyl groups attached directly to carbon atoms (C-CH$_3$). This "N-methyl" group gives rise to low wavenumber C-H stretches that are useful in identifying it. A good example is the spectrum of N-methylcyclohexylamine, which is seen in Figure 5.14. There is a low wavenumber, medium intensity C-H stretch at 2784 cm^{-1} that is easily seen. It is due to the symmetric stretch of the N-methyl group. For saturated amines in general, the N-methyl symmetric C-H stretch occurs from 2805 to 2780 cm^{-1}. For aromatic amines where a methyl group is attached to the nitrogen, this band appears from 2820 to 2810 cm^{-1}.

The N,N-dimethyl group consists of two methyl groups attached to an amine nitrogen. This group also gives rise to easily seen, low wavenumber symmetric C-H stretching bands. For a saturated N,N-dimethyl amine, two bands are found, one from 2825 to 2810 cm^{-1} and the second from 2775 to 2765 cm^{-1}. Aromatic amines with two methyl groups attached (N,N-dimethyl anilines) exhibit a low wavenumber methyl symmetric stretch from 2810 to 2790 cm^{-1}.

Although in theory CH$_2$ groups attached to a nitrogen could have low wavenumber C-H stretches, the bands tend to be weak and appear in a wider range than the N-methyl C-H stretches. This makes these N-methylene C-H stretches poor group wavenumbers. Table 5.7 summarizes the group wavenumbers for N-methyl C-H symmetric stretches.

Table 5.7 N-Methyl Symmetric Stretches (all numbers in cm^{-1}).

Grouping	Saturated	Aromatic
N-CH$_3$	2805-2780	2820-2810
N-(CH$_3$)$_2$	2825-2810, 2775-2765	2810-2790

V. Nitriles: The C≡N Bond

A. Structure and Nomenclature

A *nitrile* contains a carbon nitrogen triple bond, C≡N. The skeletal framework of a nitrile is seen in Figure 5.16. Note that the carbon involved in the triple bond is called the nitrile carbon and that a nitrile also has a carbon atom (the alpha carbon) singly bonded to the nitrile carbon.

Alpha Carbon   C—C≡N   *Nitrile Carbon*

Figure 5.16 The skeletal framework of the nitrile functional group.

If the alpha carbon of a nitrile is saturated, the compound is called a saturated nitrile. If the alpha carbon is part of a benzene ring, the molecule is denoted as an aromatic nitrile. Saturated and aromatic nitriles can be distinguished by their infrared spectra as discussed below.

Like many of the functional groups mentioned in Chapters 4 and 5, nitriles can be considered derivatives of carboxylic acids. Nitriles are named by dropping the "ic" suffix from the name of the acid from which the nitrile is made, and adding the suffix "nitrile." Thus, the nitrile made from benzoic acid is named benzonitrile. The C≡N bond also occurs in inorganic molecules, which are called cyanides. Organic nitriles are sometimes called "cyano" compounds as a result.

B. Spectroscopy of the C≡N Bond

The carbon nitrogen triple bond, C≡N, gives rise to an intense band in a unique wavenumber range, and is a classic example of a good group wavenumber. This useful infrared band is caused by the stretching of the C≡N bond. For a saturated nitrile, the C≡N stretch appears from 2260 to 2240 cm^{-1}. An example of a this band is seen at 2252 cm^{-1} in the spectrum of acetonitrile, seen in Figure 5.17.

142 Organic Nitrogen Compounds

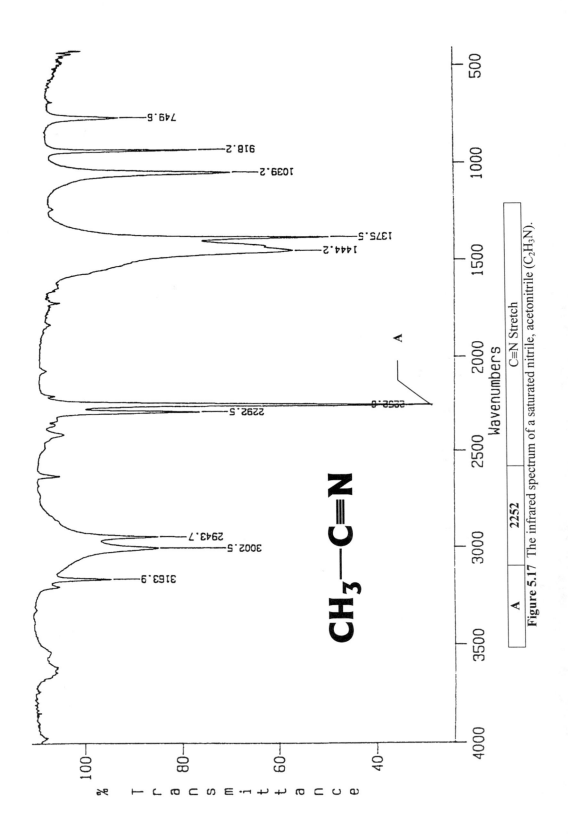

Figure 5.17 The infrared spectrum of a saturated nitrile, acetonitrile (C_2H_3N).

For an aromatic nitrile, the C≡N stretch appears from 2240 to 2220 cm^{-1}. It is lower than in saturated nitriles because of conjugation. As discussed in Chapter 4, conjugation involves interactions between the "p" electrons of the nitrile group and the "p" electrons of an aromatic ring. The orbitals for these electrons are next to each other, pointing above and below the plane of the molecule. A small amount of orbital overlap allows the electrons from the benzene ring and the C≡N to spread out across both functional groups. This lowers the force constant of the C≡N bond, causing the position of the C≡N stretch of an aromatic nitrile to be lowered about 10 to 20 cm^{-1} compared to a saturated nitrile.

Note that the position of the C≡N stretch enables saturated and aromatic nitriles to be distinguished. The position of the C≡N stretching vibration is not sensitive to branching on the alpha carbon or to the nature of the substituents on an attached benzene ring. Therefore, other types of discrimination between nitriles are not possible. The C≡N stretching positions for nitriles are summarized in Table 5.8.

Table 5.8 The C≡N Stretches of Nitriles (all numbers in cm^{-1}).

Vibration	Saturated	Aromatic
C≡N Stretch	2260-2240	2240-2220

VI. The Nitro Group

A. Structure and Nomenclature

A *nitro* molecule contains the NO$_2$ group and its structure is seen in Figure 5.18. Organic nitro molecules have one carbon atom attached to the nitro group by a C-N single bond. The chemical bonding of the NO$_2$ group is rather unusual. Normal valence considerations demand that oxygen have two bonds and nitrogen have three bonds. However, there is also a C-N bond in the nitro group and, if each oxygen is doubly bonded to the nitrogen, this would give the nitrogen five bonds total. In reality, the nitrogen/oxygen bonds in the nitro group are termed "bonds and a half" similar to the carboxylate C-O "bonds and a half" discussed in Chapter 4. The electrons in the NO$_2$ bonds are spread out across both N-O linkages. This makes the two N-O bonds equivalent.

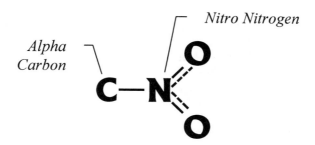

Figure 5.18 The structural framework of the nitro functional group. The dashed line denotes a nitrogen-oxygen "bond and a half."

Note that the nitrogen in the NO$_2$ group is called the *nitro nitrogen*, and that the carbon atom singly bonded to the nitro nitrogen is called the alpha carbon. Depending on whether or not the alpha carbon is saturated, or part of an aromatic ring, nitro molecules can be divided into saturated and aromatic nitro compounds.

Nitro molecules are named by simply adding the word "nitro" to the beginning of the name of the molecule that is substituted with a nitro group. For example, attaching three nitro groups to a toluene molecule results in trinitrotoluene (otherwise known as TNT), a well-known explosive.

B. The Spectroscopy of the NO_2 Group

Saturated and aromatic nitro molecules can be distinguished from their infrared spectra. Whereas aromatic nitro compounds are commonly found as chemical intermediates in organic chemistry, saturated nitro compounds are relatively rare (they are not stable). Therefore, we will limit our discussion to aromatic nitro groups. The N-O "bonds and a half" of the nitro group give rise to two intense infrared bands that are easy to identify. These bands are due to the asymmetric and symmetric stretching of the N-O bonds, which are illustrated in Figure 5.19.

Figure 5.19 The asymmetric and symmetric N-O "bond and a half" stretches of the nitro functional group.

The asymmetric stretching of an aromatic nitro group gives rise to a very strong band between 1550 and 1500 cm^{-1}. The spectrum of m-nitrotoluene in Figure 5.20 has an example of this band at 1527 cm^{-1}. Note that this band is the most intense in the spectrum. Very few other functional groups have an intense band near 1500 cm^{-1}. The symmetric stretch of the NO_2 functional group gives rise to a medium to strong intensity band from 1390 to 1330 cm^{-1}. It is seen in the spectrum of m-nitrotoluene at 1350 cm^{-1}. There are other functional groups that absorb in this region, but the intensity of this band, and the fact that it always appears along with the strong asymmetric stretch near 1500 cm^{-1}, makes it easy to identify as the nitro functional group.

The nitro group is capable of bending in a number of different directions. These vibrations give rise to several variable intensity bands at low wavenumber, so they are not good group wavenumbers. However, the NO_2 scissors vibration can sometimes be spotted. This vibration involves the O-N-O bond angle getting bigger and smaller, like the opening and closing of a pair of scissors. This vibration gives a band from 890 to 835 cm^{-1}. It is seen in Figure 5.20 at 881 cm^{-1}. The problem with this vibration is that it falls in the same range as the out-of-plane C-H bending vibrations of the aromatic ring, which can lead to some confusion.

The good news about the nitro group is that it has two strong infrared bands that are easy to spot. The bad news is that whenever the nitro group is attached to a benzene ring, it makes it impossible to determine the substitution pattern on the benzene ring. You may recall from Chapter 2 that the benzene ring out-of-plane C-H bend band, in combination with the presence or absence of the aromatic ring-bending band at 690 cm^{-1}, can be used to determine the substitution pattern on a benzene ring. The presence of a nitro group makes these rules meaningless. This is caused by the unique electronic structure of the nitro group and how it interacts electronically with the benzene ring. Suffice it to say, one must use a different analytical technique, such as NMR, to determine the substitution pattern on nitro substituted aromatic rings.

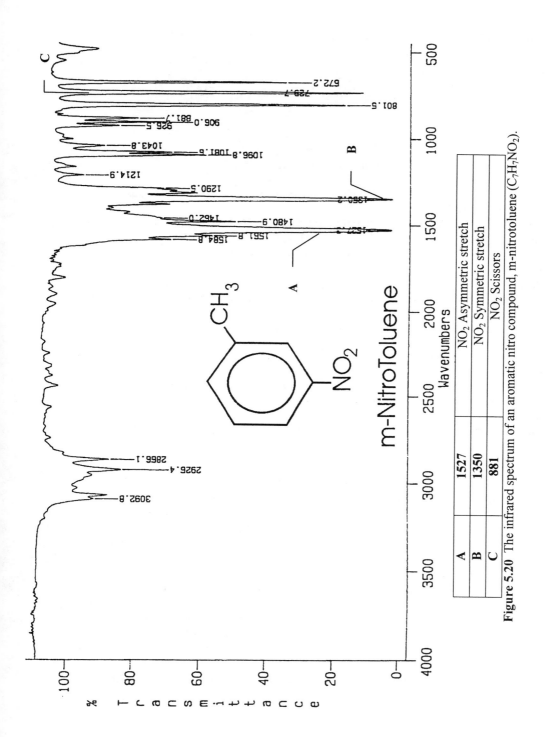

Figure 5.20 The infrared spectrum of an aromatic nitro compound, m-nitrotoluene ($C_7H_7NO_2$).

A	1527	NO_2 Asymmetric stretch
B	1350	NO_2 Symmetric stretch
C	881	NO_2 Scissors

In summary, the nitro group has two intense infrared bands due to the asymmetric and symmetric stretching of the NO₂ group. These two bands must always appear together for a sample to contain a nitro compound. The group wavenumbers for the nitro group are listed in Table 5.9.

Table 5.9 The Group Wavenumbers of the Nitro Functional Group (all numbers in cm^{-1}).

Vibration	Aromatic
NO₂ Asymmetric stretch	1550-1500
NO₂ Symmetric stretch	1390-1330
NO₂ Scissors	890-835

Bibliography

L.J. Bellamy, *The Infrared Spectra of Complex Molecules, Third Edition*, Wiley, New York, 1975.
R. Silverstein, G. Bassler, T. Morrill, *Spectrometric Identification of Organic Compounds, Fourth Edition*, Wiley, New York, 1981.
D. Lin-Vien, N. Colthup, W. Fately, J. Grasselli, *Infrared and Raman Characteristic Frequencies of Organic Molecules*, Academic Press, Boston, 1991.
N. Colthup, L. Daly, S. Wiberley, *Introduction to Infrared and Raman Spectroscopy*, Academic Press, New York, 1990.
F. Miller, *Lecture Notes for Bowdoin College Infrared Short Course*, Brunswick, Maine, 1992.
B.C. Smith, *Fundamentals of Fourier Transform Infrared Spectroscopy*, CRC Press, Boca Raton, 1996.
H. Mantsch, D. Chapman, Eds., *Infrared Spectroscopy of Biomolecules*, Wiley, New York, 1996.
A. Streitweiser, C. Heathcock, *Introduction to Organic Chemistry*, Macmillan, New York, 1976.
S. Budavari, Editor, *The Merck Index, 11th Edition*, Merck & Co., Rahway, NJ, 1989.
R.C. Weast, Editor, *The CRC Handbook of Chemistry and Physics*, CRC Press, Boca Raton, 1987.
J. Dean, Editor, *Lange's Handbook of Chemistry*, McGraw-Hill, New York, 1979.

Problem Spectra

The following spectra are presented as an exercise for the reader. Using the techniques and knowledge obtained in this chapter, do your best to predict the complete molecular structure of each sample from its infrared spectrum. The exact positions of prominent peaks are contained in the table accompanying each spectrum. However, these peaks may or may not be useful in the final determination of a structure. The physical state of the sample is also stated (solid or liquid) along with the sampling method used.

All problem set spectra were obtained on modern day FTIR instruments at 4 or 8 cm^{-1} resolution and are of pure materials. It is much easier to learn the basics of interpretation by looking at the simple spectra of pure compounds than to tackle the much more complex problem of analyzing mixture spectra. Please realize that the determination of a complete molecular structure in each problem may not be possible. However, by analyzing each of these spectra methodically and patiently, you will be able to learn a lot from each of them. The correct structure and a discussion of the interpretation of each problem are included in an appendix at the end of the book. Good Luck!

147

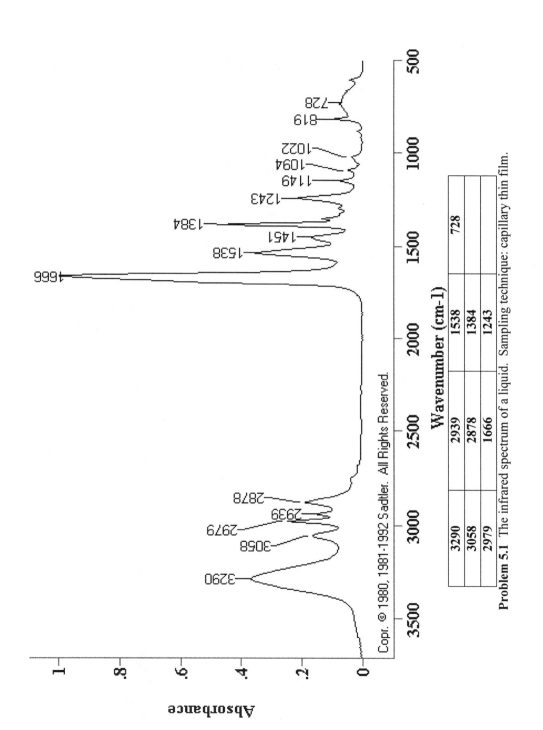

3290	2939	1538	728
3058	2878	1384	
2979	1666	1243	

Problem 5.1 The infrared spectrum of a liquid. Sampling technique: capillary thin film.

148 Organic Nitrogen Compounds

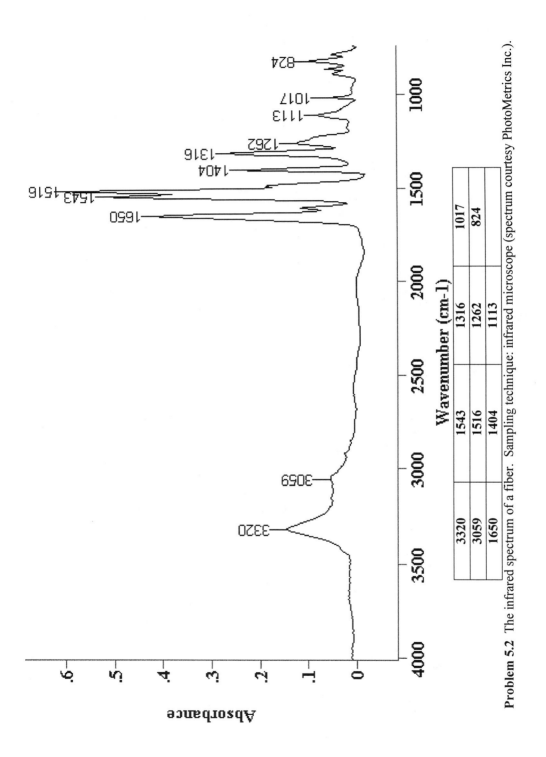

Problem 5.2 The infrared spectrum of a fiber. Sampling technique: infrared microscope (spectrum courtesy PhotoMetrics Inc.).

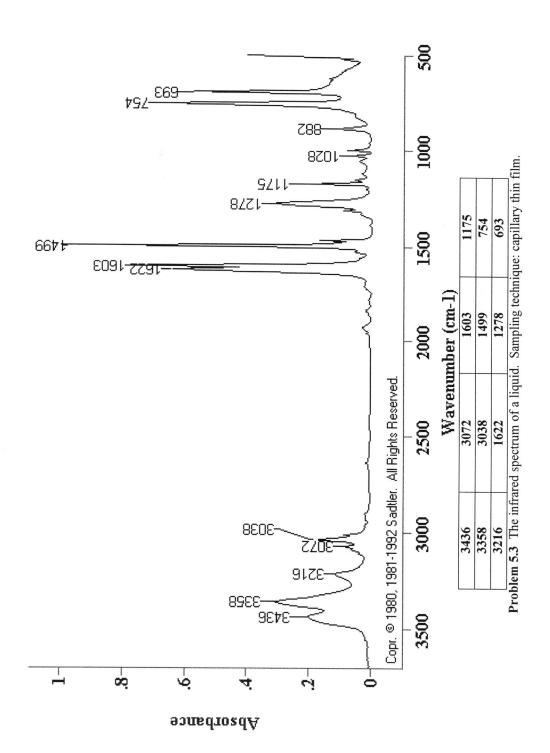

Problem 5.3 The infrared spectrum of a liquid. Sampling technique: capillary thin film.

3436	3072	1603	1175
3358	3038	1499	754
3216	1622	1278	693

150 Organic Nitrogen Compounds

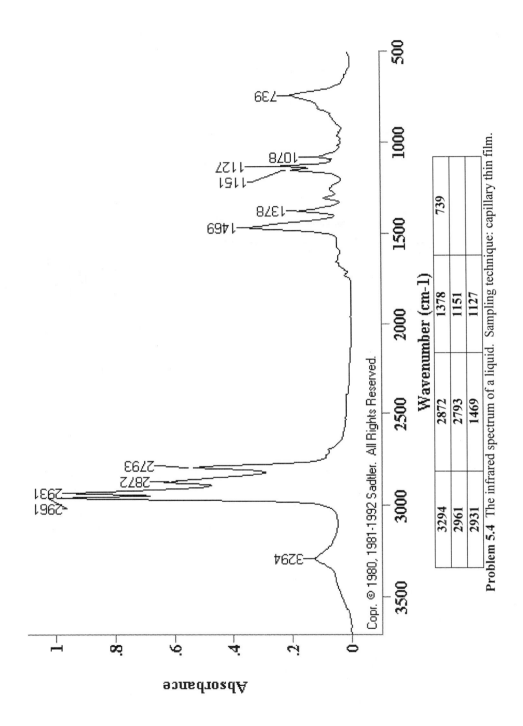

3294	2872	1378	739
2961	2793	1151	
2931	1469	1127	

Problem 5.4 The infrared spectrum of a liquid. Sampling technique: capillary thin film.

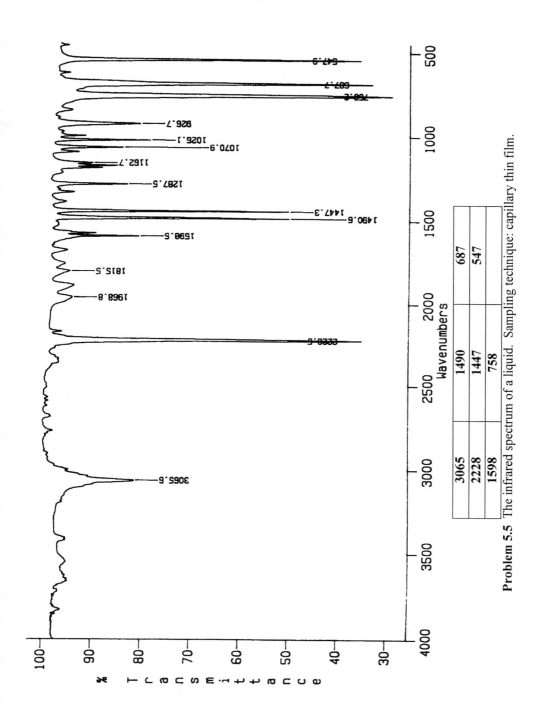

Problem 5.5 The infrared spectrum of a liquid. Sampling technique: capillary thin film.

3065	1490	687
2228	1447	547
1598	758	

152 Organic Nitrogen Compounds

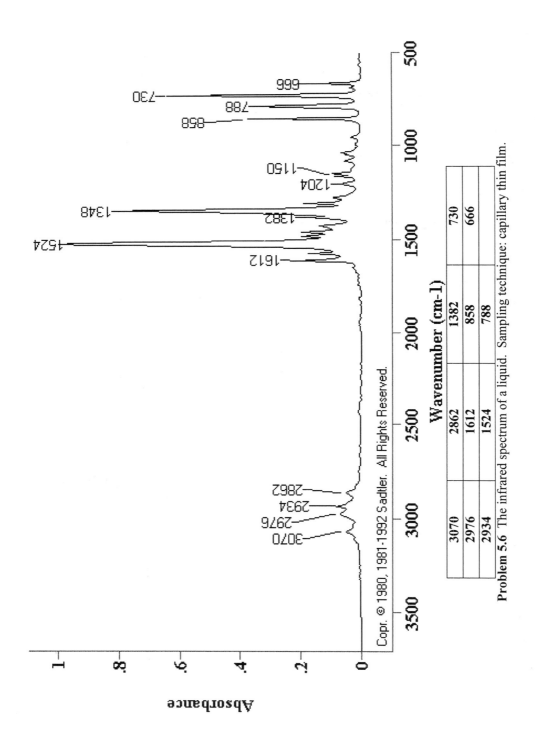

Problem 5.6 The infrared spectrum of a liquid. Sampling technique: capillary thin film.

3070	2862	1382	730
2976	1612	858	666
2934	1524	788	

Chapter 6

Organic Compounds Containing Sulfur, Silicon, and Halogens

The first five chapters of this book have focused on molecules that contain the elements carbon, hydrogen, oxygen, and nitrogen. The vast majority of organic molecules contain some or all of these chemical elements. However, sulfur, silicon, and the halogens are also found in organic molecules. The purpose of this chapter is to discuss the infrared spectra of these molecules.

I. Organic Sulfur Compounds

There are a variety of organic functional groups that contain sulfur. Sulfur is commonly found bonded to one, two, three, or four oxygen atoms, and each of these structural units gives rise to a set of functional groups. Sulfur-oxygen bonds have intense infrared bands because of their large dipole moments. However, S-S, S-H, and C-S bonds have weak infrared bands because of their small (or nonexistent) dipole moments, and are sometimes difficult to detect via infrared spectroscopy. In these cases, alternative forms of analysis may be necessary to determine if there is an S-S or S-H bond in a sample. Only the sulfur-containing functional groups with useful group wavenumbers will be discussed in this section.

A. Thiols (Mercaptans)
1. Structure and Nomenclature

The *thiol* functional group contains S-H and C-S single bonds. These molecules are sometimes called *mercaptans*. The structural framework of a thiol is shown in Figure 6.1.

$$C-S-H$$

Figure 6.1 The skeletal framework of the thiol functional group.

Thiols are the "sulfur analog" of alcohols, with the alcohol oxygen being replaced by a sulfur atom. However, the spectroscopy of thiols is totally different from that of alcohols. The S-H bond is weakly polar, so the hydrogen bonding between thiol molecules is weak. Therefore, thiol infrared bands are much narrower and weaker than alcohol infrared bands.

Thiol nomenclature involves taking the name of the hydrocarbon chain attached to the sulfur, and adding the suffix -thiol. Thus, ethanethiol has the structure shown in Figure 6.2.

$$CH_3CH_2-S-H$$

Figure 6.2 The chemical structure of ethanethiol.

154 Sulfur, Silicon, and Halogen Compounds

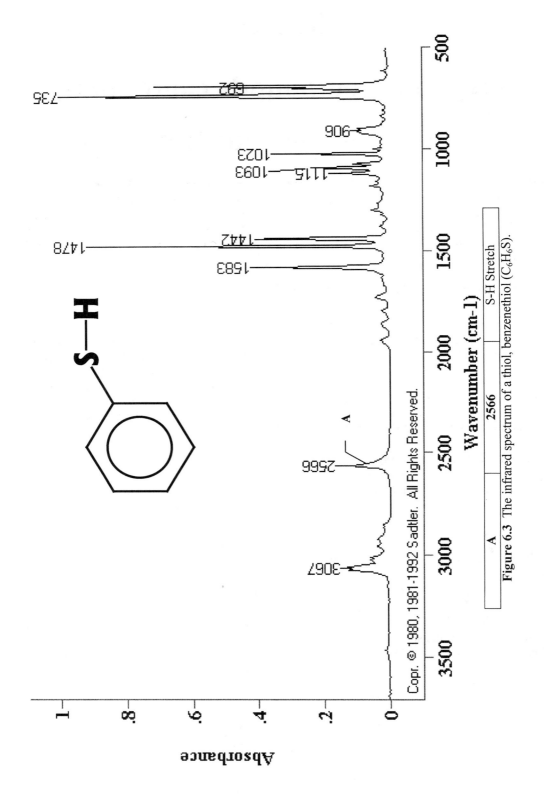

Figure 6.3 The infrared spectrum of a thiol, benzenethiol (C_6H_6S).

| A | 2566 | S-H Stretch |

One of the disagreeable properties of thiols is their strong odor. Thiols are responsible for the smell of skunks and onions. The human nose is more sensitive to thiols than any other type of chemical. Your nose can detect ethanethiol in the parts-per-billion (ppb) range, better than any laboratory instrument. Often the smell of a sample may be enough to indicate that it contains a thiol.

2. Infrared Spectra of Thiols

The infrared bands of thiols are usually medium to weak in intensity because of the small dipole moments of C-S and S-H bonds. Hydrogen bonding does occur in thiols, but it is weak compared to molecules with N-H and O-H bonds. However, the weak hydrogen bonding in thiols does cause infrared bands due to the S-H moiety to be somewhat broader than bands due to C-H and C-C bonds.

The only useful group wavenumber for thiols is the S-H stretching vibration, which is found between 2590 and 2560 cm^{-1} in the liquid phase. The S-H stretch of benzenethiol is seen in Figure 6.3 at 2566 cm^{-1}. Note that this band is medium to weak in intensity, and is slightly broader than some of the aromatic ring modes seen in this spectrum. Thiol S-H stretches appear in a unique wavenumber range, so they are usually easy to see despite their low intensity. Table 6.1 summarizes the group wavenumber information for thiols.

Table 6.1 The Group Wavenumber for Thiols (all numbers in cm^{-1}).

Vibration	Wavenumber Range
S-H Stretch	2590-2560

B. Molecules Containing Sulfur/Oxygen Bonds
1. Structure and Nomenclature

A molecule that contains an S=O double bond and two C-S single bonds is called a *sulfoxide*. The structure of a common sulfoxide is shown in Figure 6.4.

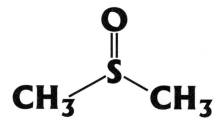

Figure 6.4 The chemical structure of dimethyl sulfoxide, a common organic solvent.

Note that a sulfoxide is the sulfur analog of a ketone, where a sulfur atom replaces the carbonyl carbon. The most important infrared band of sulfoxides, and any molecule that contains an S=O bond, is the S=O stretch. It is strong and easy to see, like the C=O stretch of carbonyl-containing functional groups.

Sulfoxides are named like ketones. The names of the two hydrocarbon substituents attached to the sulfur are listed, followed by the word "sulfoxide." Thus, the compound in Figure 6.4 is properly named dimethyl sulfoxide. There are a plethora of functional groups containing S-O and S=O bonds. The names and skeletal frameworks of several of the more common sulfur/oxygen functional groups are shown in Figure 6.5.

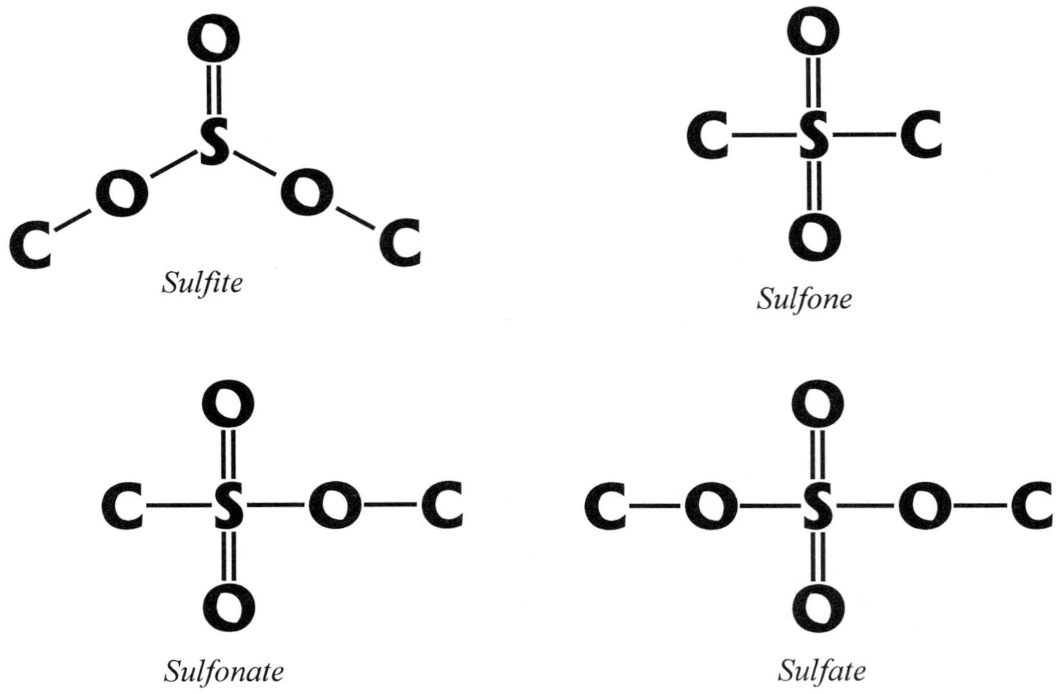

156   Sulfur, Silicon, and Halogen Compounds

Sulfite

Sulfone

Sulfonate

Sulfate

Figure 6.5 The skeletal frameworks of the sulfite, sulfone, sulfonate, and sulfate functional groups.

These compounds are named by taking the hydrocarbon substituents and adding the word sulfite, sulfone, sulfonate, or sulfate. Thus, a sulfone with two methyl groups attached to the sulfur would be dimethyl sulfone.

2. Spectra of Compounds Containing One S=O Bond

The two types of molecules with a single S=O bond that will be discussed here are sulfoxides and sulfites, whose structures can be seen in Figures 6.4 and 6.5, respectively. Note that sulfoxides have two C-S bonds, while sulfites have two S-O bonds. The infrared spectrum of the most common sulfoxide, dimethyl sulfoxide (DMSO), is seen in Figure 6.6. The most useful group wavenumber for the sulfoxide group is the S=O stretch, which appears quite strongly from 1070 to 1030 cm^{-1}. In the spectrum of DMSO, the S=O stretch appears at 1053 cm^{-1}. This band owes its intensity to the large dipole moment of the S=O bond. Note that in the spectrum of DMSO, the methyl group C-H stretching and bending bands fall into different regions than if the CH$_3$ were attached to a carbon atom.

Sulfites also contain a single S=O bond, and have a strong S=O vibration that appears from 1240 to 1180 cm^{-1}. This is higher in wavenumber than the S=O stretch of sulfoxides. It is commonly found that the wavenumber of S=O stretches increases as more oxygens are added to a sulfur atom. Table 6.2 summarizes the group wavenumbers for molecules with a single S=O bond.

Table 6.2 The Group Wavenumbers for Compounds with a Single S=O Bond

Functional Group	S=O Stretch (cm^{-1})
Sulfoxide	1070-1030
Sulfite	1240-1180

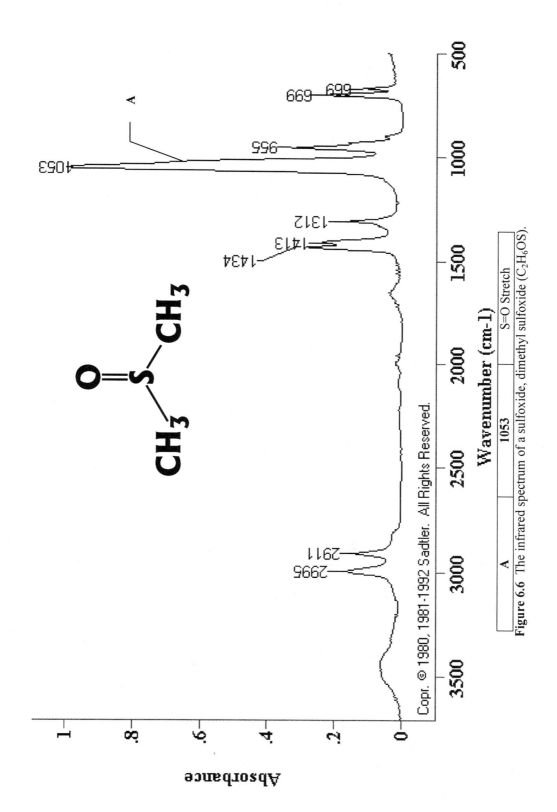

Figure 6.6 The infrared spectrum of a sulfoxide, dimethyl sulfoxide (C_2H_6OS).

| A | 1053 | S=O Stretch |

158 Sulfur, Silicon, and Halogen Compounds

3. Spectra of Compounds with Two S=O Bonds

A sulfur atom with two S=O bonds is often written in chemical shorthand as SO_2. The functional groups that contain the SO_2 group that will be discussed here are sulfones, sulfonates, and sulfates. The skeletal frameworks of these functional groups are seen in Figure 6.5. These three functional groups differ in the number of S-O bonds, containing zero, one, and two of these bonds, respectively. Because it has two S=O bonds, the SO_2 group has asymmetric and symmetric stretching vibrations. These give rise to two strong infrared bands between about 1450 and 1150 cm^{-1}. The wavenumber of these S=O stretches increase as the number of S-O bonds increases. A sample does not contain an SO_2 group unless both of these strong bands are present.

Sulfones contain two C-S bonds in addition to the SO_2 group. The asymmetric and symmetric stretches of the SO_2 group in sulfones occur at 1340 to 1310 cm^{-1} and at 1165 to 1135 cm^{-1}. Sulfonates contain one S-O bond in addition to the SO_2 group. The asymmetric and symmetric SO_2 stretches of this group appear from 1430 to 1330 cm^{-1} and 1200 to 1150, respectively. Last, sulfates contain two S-O bonds in addition to the SO_2 group. Its SO_2 stretches are found at 1450 to 1350 cm^{-1} and 1230 to 1150 cm^{-1}. Table 6.3 summarizes the asymmetric and symmetric SO_2 stretching positions for these functional groups. The infrared spectra of inorganic sulfates are discussed in the next chapter.

Table 6.3 The Group Wavenumbers for SO_2 Containing Molecules (all numbers in cm^{-1})

Functional Group	Asymm. SO_2 Stretch	Sym. SO_2 Stretch
Sulfone	1340-1310	1165-1135
Sulfonate	1430-1330	1200-1150
Sulfate	1450-1350	1230-1150

II. Organic Silicon Compounds

A. Structure and Nomenclature of Siloxanes (Silicones)

The most commonly encountered organic molecules that contain silicon atoms are called *siloxanes*. A siloxane contains a chain of Si-O-Si bonds, with organic functional groups attached to the Si atoms. The chemical structure of a very common siloxane, polydimethylsiloxane, is shown in Figure 6.7.

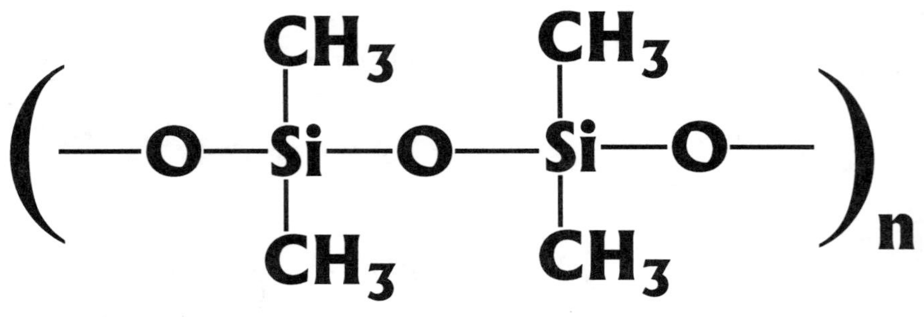

Figure 6.7 The chemical structure of polydimethylsiloxane.

Siloxanes are also known by their common name of *silicones*. Silicones tend to be oils and greases at room temperature, which makes them very useful as lubricants. Additionally, silicone sprays are commonly used to waterproof clothing, shoes, and furniture. Siloxanes are named by appending the word "siloxane" to the names of the organic moieties attached to the silicon atom. Thus, the compound in Figure 6.7 is an example of a dimethylsiloxane.

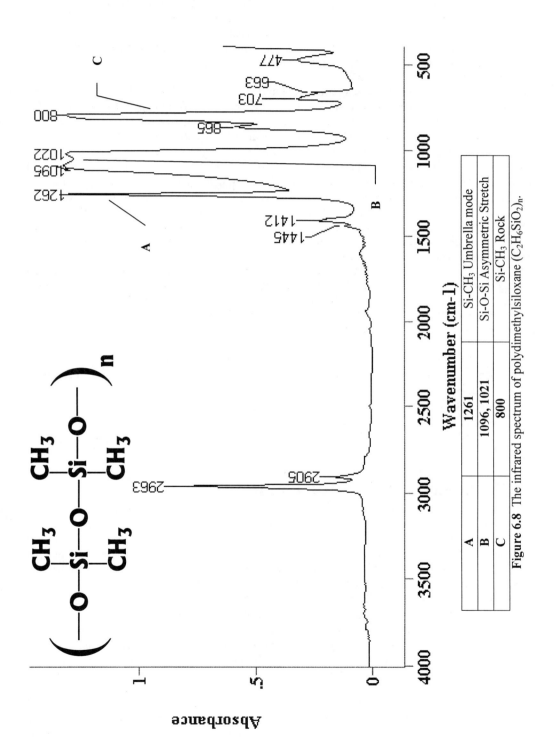

Figure 6.8 The infrared spectrum of polydimethylsiloxane $(C_2H_6SiO_2)_n$.

A	1261	Si-CH$_3$ Umbrella mode
B	1096, 1021	Si-O-Si Asymmetric Stretch
C	800	Si-CH$_3$ Rock

159

B. Spectra of Siloxanes (Silicones)

Although the element silicon is in the same chemical family as carbon, and commonly forms 4 chemical bonds, it has a much lower electronegativity than carbon. Therefore, silicon forms polar bonds with oxygen and carbon atoms, causing the infrared bands due to these functional groups to be more intense than in molecules that do not contain Si atoms. The infrared spectrum of a very common siloxane, polydimethylsiloxane, is seen in Figure 6.8. All siloxanes have a Si-O-Si chain in common. The asymmetric stretch of the bonds in this chain gives rise to one or two very intense bands between 1130 and 1000 cm^{-1}. These bands appear in Figure 6.8 at 1081 and 1017 cm^{-1}.

Methyl groups attached to a silicon atom will undergo the same C-H stretching and bending vibrations as a CH_3 attached to a carbon atom. However, the positions of the bands for a Si-CH_3 group will be different than for a C-CH_3 group, because of changes electronic effects. The best group wavenumber for the Si-CH_3 group is the umbrella mode (symmetric bend), which is a very intense band at 1260±5 cm^{-1}. This band appears in Figure 6.8 at 1259 cm^{-1}. If a silicon atom has two methyl groups attached to it, denoted as $Si(CH_3)_2$, there is a strong methyl rocking mode band that appears at 800±10 cm^{-1}. This band is seen in the spectrum of polydimethylsiloxane at 800 cm^{-1}. Polydimethylsiloxane is the main component of stopcock grease, and is one of the most commonly found contaminants in organic samples. The pattern of bands in its spectrum is very characteristic; a series of 4 intense bands between 1200 and 800 cm^{-1}. Few other materials give rise to this pattern, and since polydimethylsiloxane is such a common contaminant, this pattern of bands is worth remembering.

If a silicon atom has three methyl groups attached to it, denoted as $Si(CH_3)_3$, there are two rocking modes seen at ~840 and ~760 cm^{-1}. If a hydrocarbon chain with two or more carbons is attached to a Si atom, then a Si-CH_2 bond will exist. This structural unit gives rise to a CH_2 wag band from 1250 to 1200 cm^{-1} and a CH_2 rock from 760 to 670 cm^{-1}. Table 6.4 summarizes the group wavenumbers for siloxanes.

Table 6.4 The Group Wavenumbers of Siloxanes (all numbers in cm^{-1})

Functional Group	CH$_3$ Umbrella Mode	CH$_3$ Rock	CH$_2$ Wag and Rock
Si-CH$_3$	1260± 5	-	-
Si(CH3)$_2$	1260± 5	800±10	-
Si(CH$_3$)$_3$	1260± 5	~840, ~760	-
Si-CH$_2$	-	-	1250-1200, 760-670

III. Halogenated Organic Compounds

The halogens are a family of chemical elements that comprise an entire column of the periodic table. The most commonly found halogens in organic compounds are fluorine, chlorine, bromine, and iodine. The halogens are very electronegative, forming polar bonds with carbon atoms with large dipole moments. Therefore, carbon halogen stretches, or C-X stretches where X represents a halogen, are very intense. Since halogens are generally heavier than the other atoms found in organic molecules, C-X stretches tend to appear at low wavenumber (a mass effect). The combination of high intensity and low wavenumber makes C-X stretches relatively easy to spot in an infrared spectrum. In halogenated compounds with more than one carbon atom, rotational isomers (rotamers) are possible, leading to the appearance of several bands in the C-X stretching region. This can make assigning C-X stretches difficult. The high electronegativity of halogens introduces electronic effects into the band positions of C-Hs bonded to the same carbon as the halogen. The positions and intensities of these C-H stretches will be totally different than found in other organic molecules, and are not good group wavenumbers. Some aromatic ring bands appear in the same region as C-X stretches. When a halogen is attached to an aromatic ring, it can be difficult to determine whether or not a band is due to the ring or the C-X stretch. Consequently, there are no good group wavenumbers for halogenated aromatic rings.

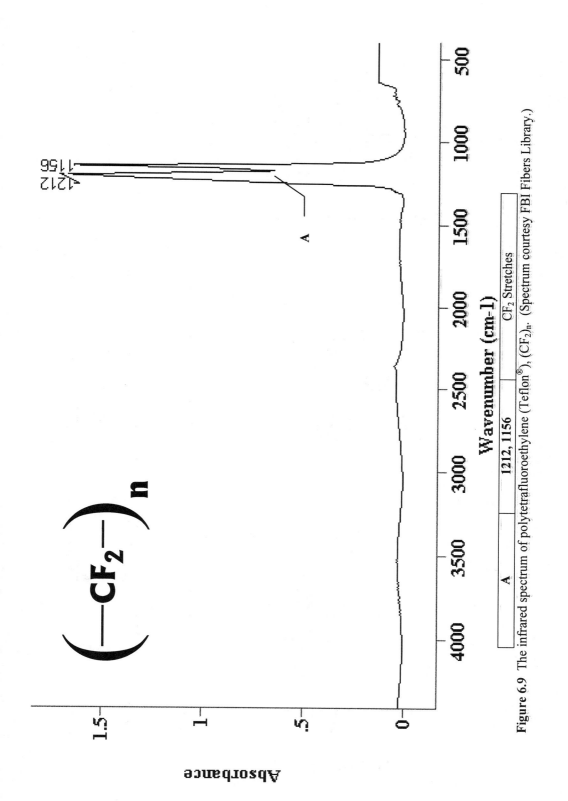

Figure 6.9 The infrared spectrum of polytetrafluoroethylene (Teflon®), $(CF_2)_n$. (Spectrum courtesy FBI Fibers Library.)

162 Sulfur, Silicon, and Halogen Compounds

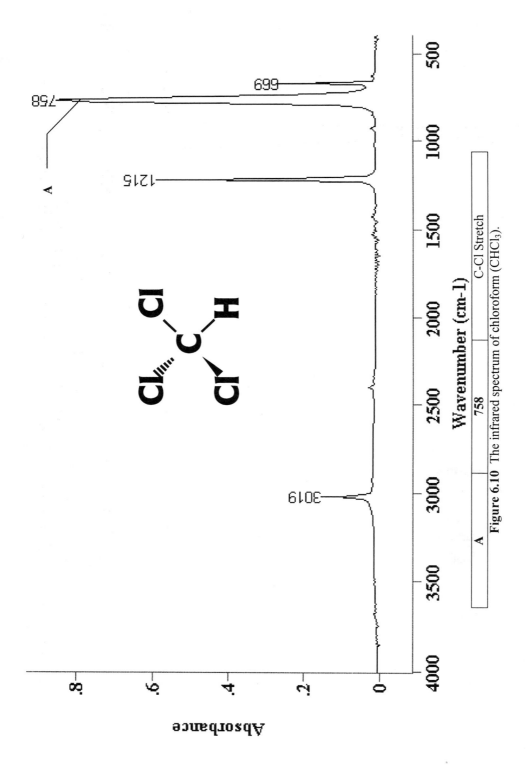

Figure 6.10 The infrared spectrum of chloroform (CHCl$_3$).

| A | 758 | C-Cl Stretch |

The fluorine atom shows the highest wavenumber C-X stretches since it is the lightest halogen. The infrared spectrum of polytetrafluoroethylene (Teflon®) is seen in Figure 6.9. The two strong bands at 1212 and 1156 cm^{-1} are the asymmetric and symmetric CF_2 stretches, respectively. C-F stretches generally appear from 1300 to 1000 cm^{-1}. C-F bonds also have a bending vibration at ~650 cm^{-1}. This band appears in the spectrum of Teflon, but does not appear in Figure 6.9 because the detector that was used to obtain this spectrum cut off at 700 cm^{-1}.

Chlorine is the halogen under fluorine in the periodic table of the chemical elements. Its atomic weight of 35.45 makes it almost twice as heavy as fluorine. Consequently, C-Cl stretches occur at lower wavenumber than C-F stretches (a mass effect). C-Cl stretches are typically found from 800 to 600 cm^{-1}. The infrared spectrum of chloroform in Figure 6.10 has a C-Cl stretching band at 758 cm^{-1}. Note the unusual position of the C-H stretching and bending bands in this spectrum, at 3019 and 1215 cm^{-1} respectively.

The heavier halogens, bromine and iodine, have C-X stretches at even lower wavenumber than chlorine. Bands due to C-Br stretching appear from 650 to 500 cm^{-1}, and bands due to C-I stretching appear from 570 to 500 cm^{-1}. These bands are similar in intensity to the C-Cl stretch seen in Figure 6.10. Table 6.5 summarizes the group wavenumbers for C-X stretches. Note that the general C-X stretching wavenumber range goes down with increasing atomic mass. However, the ranges of some C-X stretches overlap, sometimes making it difficult to determine what halogen is present in a sample.

Table 6.5 The C-X Stretches for Halogenated Organic Molecules

Bond	**C-X Stretch**
C-F	1300-1000
C-Cl	800-600
C-Br	650-550
C-I	570-500

Bibliography

L.J. Bellamy, *The Infrared Spectra of Complex Molecules, Third Edition*, Wiley, New York, 1975.

R. Silverstein, G. Bassler, T. Morrill, *Spectrometric Identification of Organic Compounds, Fourth Edition*, Wiley, New York, 1981.

D. Lin-Vien, N. Colthup, W. Fately, J. Grasselli, *Infrared and Raman Characteristic Frequencies of Organic Molecules*, Academic Press, Boston, 1991.

N. Colthup, L. Daly, S. Wiberley, *Introduction to Infrared and Raman Spectroscopy*, Academic Press, New York, 1990.

F. Miller and J. Graselli, *Lecture Notes for Bowdoin College Infrared Short Course*, Brunswick, Maine, 1992.

B.C. Smith, *Fundamentals of Fourier Transform Infrared Spectroscopy*, CRC Press, Boca Raton, 1996.

H. Mantsch, D. Chapman, Eds., *Infrared Spectroscopy of Biomolecules*, Wiley, New York, 1996.

A. Streitwieser, C. Heathcock, *Introduction to Organic Chemistry*, Macmillan, New York, 1976.

S. Budavari, Editor, *The Merck Index, 11th Edition*, Merck & Co., Rahway, NJ, 1989.

R.C. Weast, Editor, *The CRC Handbook of Chemistry and Physics*, CRC Press, Boca Raton, 1987.

J. Dean, Editor, *Lange's Handbook of Chemistry*, McGraw-Hill, New York, 1979.

Chapter 7

Inorganic Compounds

I. Introduction

Infrared spectroscopy is not always thought of as an inorganic analysis tool. It is true that most of the time it is organic molecules whose spectra we are interested in, but many inorganic molecules have useful infrared spectra also. There is no reason why infrared spectra of inorganic materials should not be obtained.

Many inorganic molecules consist of a positively charged metal ion bonded to a negatively charged ion (ionic bonding). Because of the charges on these ions, the dipole moment of these bonds is large. Thus, the infrared bands of inorganic molecules are characteristically strong. An inorganic molecule will have a useful group wavenumber in the mid-infrared (4000 to 400 cm^{-1}) if it contains a polyatomic ion that contains light elements such as carbon, nitrogen, oxygen, and sulfur. It is vibrations such as C-O, N-O, and S=O stretching and bending that give rise to good inorganic group wavenumbers in the mid-infrared. These vibrations are similar to the C-O, N-O, and S=O vibrations found in organic molecules. A difference between organic and inorganic molecules is that inorganic molecules do not contain C-H bonds. Thus, C-H stretching and bending bands are conspicuously absent from inorganic spectra. One if the surest ways to distinguish between organic and inorganic molecules is by the presence or absence of C-H stretching bands.

Infrared bands due to metal atoms, such as metal-oxygen stretching and bending vibrations, appear below 400 cm^{-1} because of the large mass of the metal atoms. Most infrared spectrometers do not work in this wavenumber range (although there are some that do), making it difficult to observe these bands. It is usually impossible to tell from a mid-infrared spectrum what metal atoms are present in a sample. Atomic spectroscopy instruments that are well suited to identifying the metal atoms in a sample should be used when this type of information is needed.

Most inorganic substances are found in a crystalline state. The way in which atoms are arranged in a crystal affects the infrared spectrum of a material. For example, calcium carbonate ($CaCO_3$) exists in nature as two different minerals, calcite and aragonite. These two minerals have the same chemical composition, but their atoms are arranged differently in space. Thus, they have different infrared spectra, and infrared spectra can be used to distinguish between minerals with the same chemical composition.

A characteristic of inorganic spectra is the presence of water bands. Many organic molecules are hydrophobic, so water is not found in samples of these molecules. Since inorganics contain charged bonds, the polar ends of the O-H bond in a water molecule hydrogen bond to inorganic molecules. There are several different types of water molecules present in inorganic molecules. The first are complexes in which the water molecule is held in place by a covalent bond. The second, known as waters of hydration, occupy a space in the crystal lattice of an inorganic substance, but are only weakly bound. The third is water adsorbed on the surface of inorganic materials, and held in place by chemical or physical forces. These water molecules exist in different chemical environments, and may give complicated O-H stretching and bending spectra with multiple bands.

166   Inorganic Compounds

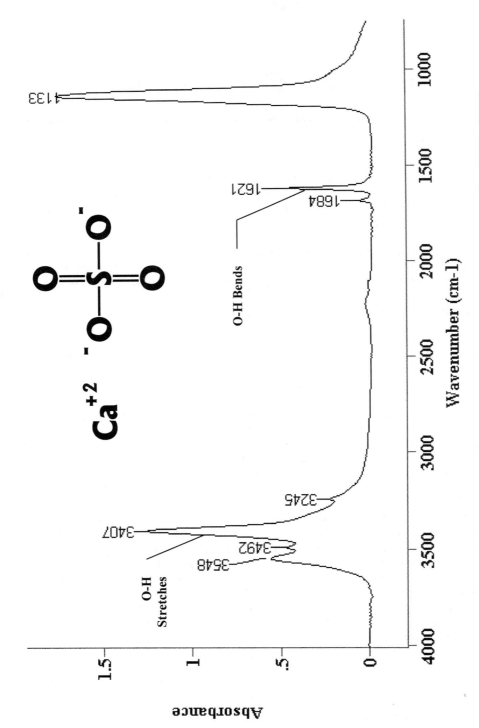

Figure 7.1 The infrared spectrum of the mineral gypsum, also known as calcium sulfate ($CaSO_4 \cdot 2H_2O$). Note the complex O-H stretching and bending bands around 3500 and 1630 cm^{-1}.

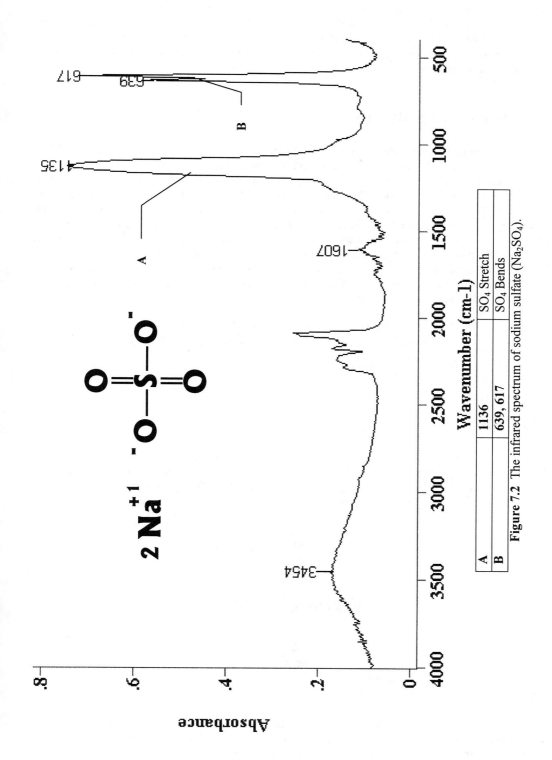

Figure 7.2 The infrared spectrum of sodium sulfate (Na$_2$SO$_4$).

A	1136	SO$_4$ Stretch
B	639, 617	SO$_4$ Bends

168 Inorganic Compounds

This is well illustrated in the spectrum of the mineral gypsum ($CaSO_4 \cdot 2H_2O$) seen in Figure 7.1. Gypsum contains two waters of hydration in addition to water adsorbed on its surface. The group of bands from 3550 to 3400 cm^{-1} and the bands at 1684 and 1621 cm^{-1} are the O-H stretches and bends of the different water molecules in gypsum. Unfortunately, there is little consistency to the width and wavenumber positions of the bands due to these different types of water molecules. This makes it difficult to distinguish between them. In summary, inorganic spectra consist of high intensity, low wavenumber vibrations characterized by the lack of C-H stretches and the presence of water bands.

II. Inorganic Sulfates

Sulfates contain the SO_4^{-2} structural unit. The structure of a sulfate is seen in Figure 7.1 Note that, nominally, a sulfate contains two S=O and two S-O bonds. Actually, the four S-O bonds are equivalent. This gives rise to one strong sulfur/oxygen stretching vibration, rather than separate S=O and S-O stretches. The sulfur-oxygen stretch of inorganic sulfates is found from 1140 to 1080 cm^{-1}. It is seen in the spectrum of sodium sulfate in Figure 7.2 at 1136 cm^{-1}. Like any other bond, sulfate bonds can bend giving rise to one or two bands in the 680 to 610 cm^{-1} range. These bands are seen in the spectrum of sodium sulfate at 639 and 617 cm^{-1}. Note that the bending bands are sharper than the stretching bands. This is commonly observed in inorganic infrared spectra. The unmarked group of peaks near 2000 cm^{-1} are overtones and combination bands of the lower wavenumber S-O stretching and bending vibrations. Table 7.1 summarizes the group wavenumbers for inorganic sulfates.

Table 7.1 The Group Wavenumbers of Inorganic Sulfates

Vibration	Wavenumber Range (cm^{-1})
S-O Stretch	1140-1080
S-O Bends	680-610

III. Silica

Silicon is in the same column of the periodic table as carbon, and forms four bonds in a tetrahedral arrangement like carbon. In inorganic compounds, the silicon atom is most frequently found bonded to four oxygen atoms as part of the SiO_4^{-4} group. The structure of the SiO_4^{-4} tetrahedron is shown in Figure 7.3.

Figure 7.3 The structure of a SiO_4^{-4} tetrahedron.

The pure form of SiO_4^{-4} is called silica, where the SiO_4^{-4} tetrahedra form long Si-O-Si chains. In its crystalline form, silica is the mineral quartz. If silica is melted and rapidly cooled, the silica tetrahedra do not have time to crystallize, and common glass is formed. Silica is used in industry as a catalyst support, polymeric filler, abrasive, adsorbent, and as a chromatographic stationary phase.

The surface chemistry of silica is important. A native silica surface will react upon exposure to atmospheric moisture to form Si-OH or *silanol* bonds. The number, position, acidity, and basicity of these silanols gives rise to the complex surface chemistry of silica. If Si-OH bonds are packed close together, they will hydrogen bond with each other. However, if the density of Si-OH bonds on the silica surface is small, lone silanol groups will result. The varied surface chemistry of silica makes it a useful adsorbent and chromatographic material. The structures of hydrogen bonded and lone silanols are seen in Figure 7.4.

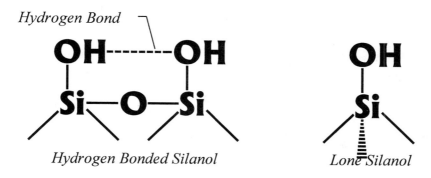

Figure 7.4 The structures of hydrogen bonded and lone silanols.

SiO_4^{-4} groups are commonly found in the earth's crust as *silicates*. Many common minerals such as clays and granites are silicates, along with certain gemstones such as emerald, sapphire, and beryl. It is the ability of the SiO_4^{-4} group to form chains, sheets, and cyclic structures that gives rise to the wide variety of minerals known to geologists.

The infrared spectrum of silica can be seen in Figure 7.5. Like many inorganic materials, silica contains large amounts of adsorbed water as seen by the broad peaks at ~3350 and 1620 cm^{-1}. The surface silanols also give rise to O-H stretching bands, which can appear from 3700 to 3200 cm^{-1}. The O-H stretching band in Figure 7.5 has some structure to it on the high wavenumber side. The shoulder at 3738 cm^{-1} is the O-H stretch of a lone silanol group. The Si-O stretch of silanols usually gives a band near 940 cm^{-1}. This band is seen in Figure 7.5 at 944 cm^{-1}.

The low wavenumber bands in Figure 7.5 are due to the Si-O-Si chain proper. The most intense band at 1085 cm^{-1} is due to asymmetric stretching of the Si-O-Si chain. This band usually appears between 1200 and 1000 cm^{-1}. Not only is this band very intense, but it has an asymmetric shape, which can be used as a diagnostic marker for the presence of silica in a sample. For example, the presence of silica in filled polymers is often revealed by the presence and shape of this band. The symmetric stretch of the Si-O-Si chain appears in Figure 7.5 at 802 cm^{-1} and is generally found at ~805cm^{-1}. The strong band at 464 cm^{-1} is due to a bending vibration of the Si-O-Si chain. Last, the two small bands near 1900 cm^{-1} are overtone and combination bands. Table 7.2 summarizes the group wavenumber information for silica and silicates.

Table 7.2 The Group Wavenumbers of Silica

Vibration	Wavenumber Range (cm^{-1})
Silanol SiO-H Stretch	3700-3200
Si-O-Si Asymmetric stretch	1200-1000
Silanol Si-O Stretch	~940
Si-O-Si Symmetric stretch	~805
Si-O-Si Bend	~450

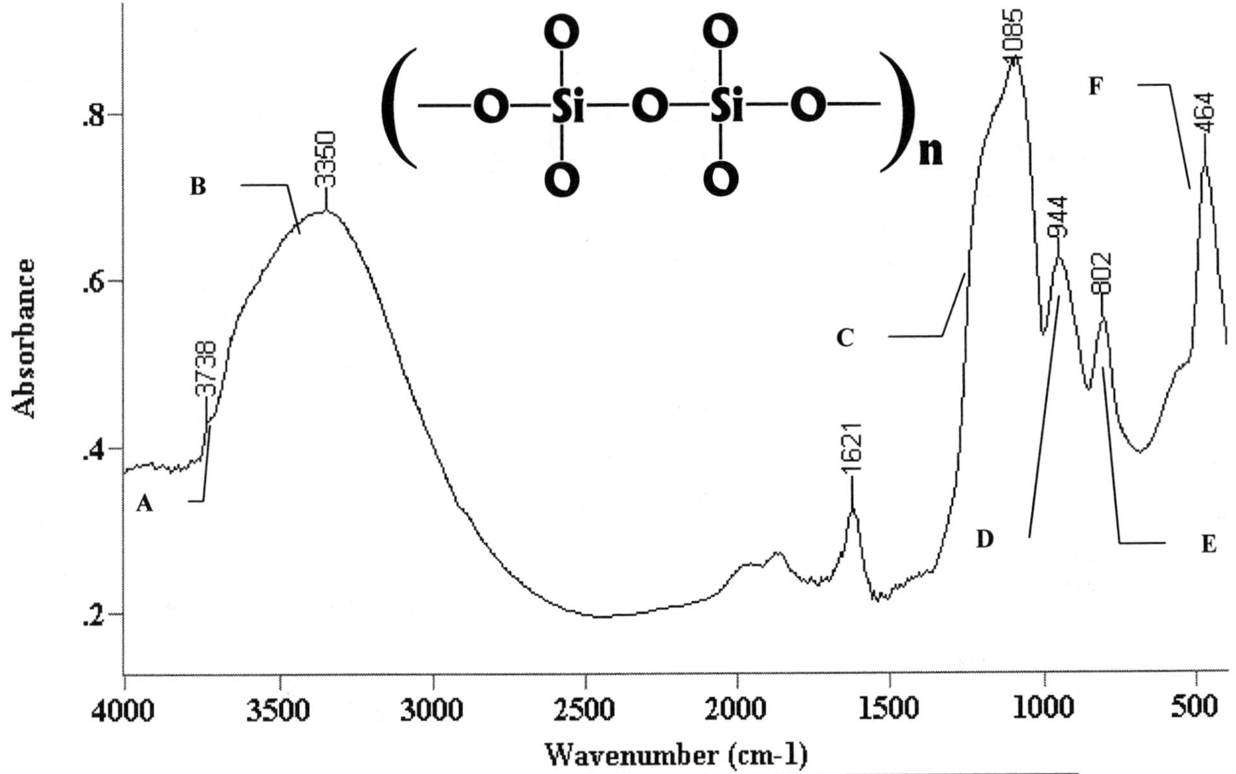

170 Inorganic Compounds

A	3738	Lone silanol SiO-H Stretch
B	3350	H-Bonded silanol and adsorbed water O-H stretch
C	1085	Si-O-Si Asymmetric stretch
D	944	Silanol Si-O stretch
E	802	Si-O-Si Symmetric stretch
F	464	Si-O-Si Bend

Figure 7.5 The infrared spectrum of silica $(SiO_4)_n$.

IV. Inorganic Carbonates

Organic and inorganic *carbonates* have the structural unit CO_3. In the organic form of carbonates, there is one C=O bond, two C-O bonds, and organic substituents attached to the singly bonded oxygens. Inorganic carbonates have a charge of minus two, and in theory have the same structure as organic carbonates. However, it is more correct to think of inorganic carbonates as having three equivalent C-O bonds arranged in a trigonal planar pattern around the carbon atom, with a bond angle of 120°. The structures of an organic and inorganic carbonate are shown in Figure 7.6. The spectra of organic carbonates are discussed in Chapter 4.

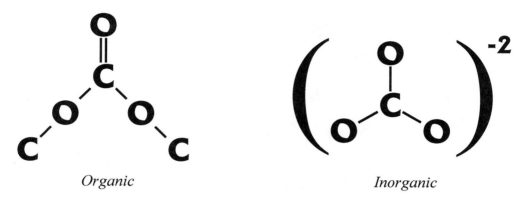

Organic   *Inorganic*

Figure 7.6 The structures of organic and inorganic carbonates.

Because of their structure, inorganic carbonates do not have separate C-O and C=O stretching vibrations. Instead, a generic "C-O" stretch is found between 1510 and 1410 cm^{-1}. This band appears in the spectrum of calcium carbonate, seen in Figure 7.7, at 1463 cm^{-1}. The position of this band is about midway between where C=O double bonds and C-O single bonds appear. This indicates that the bond order in an inorganic carbonate is somewhere between one and two.

Since an inorganic carbonate group is planar, there are out-of-plane and in-plane C-O bending vibrations. These appear from 880 to 860 cm^{-1} and at ~740 cm^{-1}, respectively. The out-of-plane bend is seen in Figure 7.7 at 877 cm^{-1}. The in-plane bend is not seen in this spectrum because the detector used to measure the spectrum cuts off at 750 cm^{-1}. Table 7.3 summarizes the group wavenumbers for inorganic carbonates.

Table 7.3 The Group Wavenumbers for Inorganic Carbonates

Vibration	Wavenumber Range (cm^{-1})
C-O Stretch	1510-1410
C-O Out-of-plane bend	880-860
C-O In-plane bend	~740

V. Nitrates

The *nitrate* group, NO_3^{-1}, is similar in structure to the carbonate group. The nitrogen-oxygen bonds lie in the same plane with a bond angle of 120°, in what is known as a trigonal planar arrangement. As with the carbonates, the three nitrogen bonds are equivalent, giving rise to a generic "nitrogen-oxygen" stretch rather than individual N-O and N=O stretches. The structure of the nitrate group is seen in Figure 7.8.

172   Inorganic Compounds

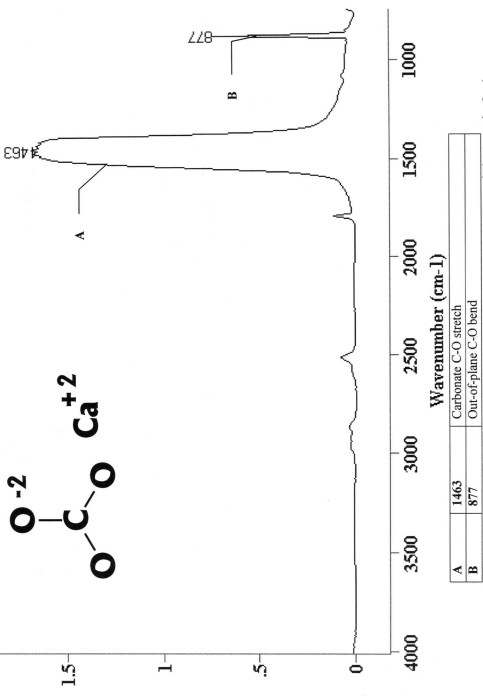

Figure 7.7 The infrared spectrum of calcium carbonate ($CaCO_3$). (Spectrum courtesy of Photometrics Inc.)

A	1463	Carbonate C-O stretch
B	877	Out-of-plane C-O bend

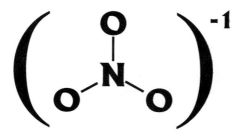

Figure 7.8 The structure of the nitrate group.

The nitrogen/oxygen stretch of nitrates appears as an intense band between 1400 and 1340 cm^{-1}. It is found in the spectrum of sodium nitrate in Figure 7.9 at 1346 cm^{-1}. The out-of-plane and in-plane bending vibrations of the nitrate group appear from 840 to 810 cm^{-1} and at ~720 cm^{-1}. These bands are seen in Figure 7.9 at 836 and 726 cm^{-1}. The medium intensity bands in Figure 7.9 above 1500 cm^{-1} are overtone and combination bands (see Chapter 1), and are not useful in identifying nitrates. Table 7.4 summarizes the group wavenumber information for nitrates.

Table 7.4 The Group Wavenumbers for Nitrates

Vibration	Wavenumber Range (cm^{-1})
N-O Stretch	1400-1340
N-O Out-of-plane bend	840-810
N-O In-plane bend	~720

VI. Phosphates

Inorganic *phosphates* are composed of a PO_4^{-3} anion and a metallic cation. There may also be different numbers of waters of hydration in a phosphate leading to a myriad of different types of molecules. The important mid-infrared vibrations of the phosphate group are phosphorous-oxygen stretching and bending vibrations. The stretch is very intense and broad, and appears between 1100 and 1000 cm^{-1}. Unfortunately, sulfates and silicates also have strong, broad bands in this region. One must make use of secondary bands, such as bending vibrations, to be able to distinguish between these types of molecules. The important secondary band for phosphates is a bending vibration found between 600 and 500 cm^{-1}. The following table summarizes the vibrational bands of phosphates.

Table 7.5 The Group Wavenumbers of Phosphates

Vibration	Wavenumber (cm^{-1})
PO_4^{-3} Stretch	1100-1000 (broad and strong)
PO_4^{-3} Bend	600-500

174   Inorganic Compounds

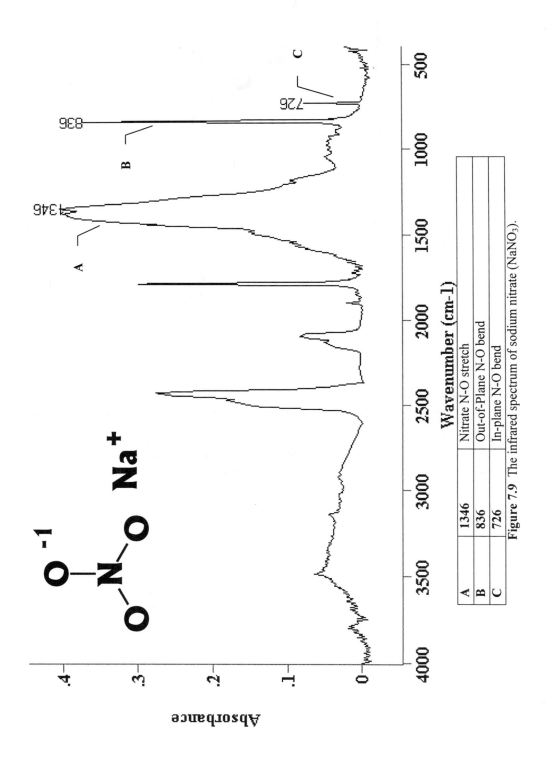

Figure 7.9 The infrared spectrum of sodium nitrate (NaNO$_3$).

A	1346	Nitrate N-O stretch
B	836	Out-of-Plane N-O bend
C	726	In-plane N-O bend

Bibliography

L.J. Bellamy, *The Infrared Spectra of Complex Molecules, Third Edition*, Wiley, New York, 1975.

R. Silverstein, G. Bassler, T. Morrill, *Spectrometric Identification of Organic Compounds, Fourth Edition*, Wiley, New York, 1981.

D. Lin-Vien, N. Colthup, W. Fately, J. Grasselli, *Infrared and Raman Characteristic Frequencies of Organic Molecules*, Academic Press, Boston, 1991.

N. Colthup, L. Daly, S. Wiberley, *Introduction to Infrared and Raman Spectroscopy*, Academic Press, New York, 1990.

F. Miller, J. Graselli, *Lecture Notes for Bowdoin College Infrared Short Course*, Brunswick, Maine, 1992.

B.C. Smith, *Fundamentals of Fourier Transform Infrared Spectroscopy*, CRC Press, Boca Raton, 1996.

H. Mantsch, D. Chapman, Eds., *Infrared Spectroscopy of Biomolecules*, Wiley, New York, 1996.

A. Streitweiser, C. Heathcock, *Introduction to Organic Chemistry*, Macmillan, New York, 1976.

S. Budavari, Editor, *The Merck Index, 11th Edition*, Merck & Co., Rahway, NJ, 1989.

R.C. Weast, Editor, *The CRC Handbook of Chemistry and Physics*, CRC Press, Boca Raton, 1987.

J. Dean, Editor, *Lange's Handbook of Chemistry*, McGraw-Hill, New York, 1979.

R. Nyquist and R. Kagel, *Infrared Spectra of Inorganic Compounds*, Academic Press, New York, 1971.

Chapter 8

Infrared Spectra of Polymers

I. Introduction

Polymer molecules consist of long chains of chemical structural units, each of which has the same structure called a *repeat unit*. For example, polyethylene is a chain of CH_2 repeat units. Polymers are made from individual molecules called *monomers*. A defining characteristic of polymer molecules is their high molecular weight. In general, infrared spectra do not respond to changes in the molecular weight of a sample. The spectrum of a methylene group is the same whether the methylene is part of a hexane molecule with a molecular weight of 86 or part of a polyethylene chain with a molecular weight of 100,000. Infrared spectra can disclose the structure of a polymeric repeat unit, but have little to say about molecular weight.

Since the spectra of polymers resemble the spectra of small molecules, why bother with a chapter on the infrared spectra of polymers? Because polymeric materials are important. Billions of pounds of polymers are made each year, and polymeric products generate billions of dollars of revenue each year for a myriad of companies. Practitioners of the art of infrared spectroscopy will be called upon to analyze a polymeric sample at some point in their career. The point of this chapter will be to act as a reference source for the spectra of some common polymers. Workers who routinely take spectra of polymeric materials should find this chapter useful in helping identify the polymers in their samples.

Almost all the functional groups discussed in this chapter have been discussed in previous chapters, and it will be assumed here that the reader is familiar with the contents of the rest of this book. In this chapter, structure and nomenclature discussions will be abandoned because this material was covered in the chapters about small molecules. Instead, the spectra of different polymers will be presented, and then the bands in each spectrum will be assigned and discussed.

II. Recyclable Plastics

This section will discuss the spectra of plastics that are commonly collected by municipalities and recycled in quantities of millions of pounds per year. These plastics are commonly used in packaging, in clothing, and as beverage containers. The materials in this class include polyethylene, polypropylene, polystyrene, and polyethylene terephthalate (PET).

A. Low Density and High Density Polyethylene

Polyethylene is an economically important polymer. Billions of pounds of it are made per year. It is present in all aspects of our daily lives, showing up as shopping bags, garbage bags, and in many other types of packaging. Polyethylene is one of the six packaging plastics that are commonly recycled. Polyethylene is made by the polymerization of ethylene ($CH_2=CH_2$), and its chemical formula is simply $[CH_2]_n$. Since it is such a simple molecule, it has a simple spectrum as seen in Figure 8.1.

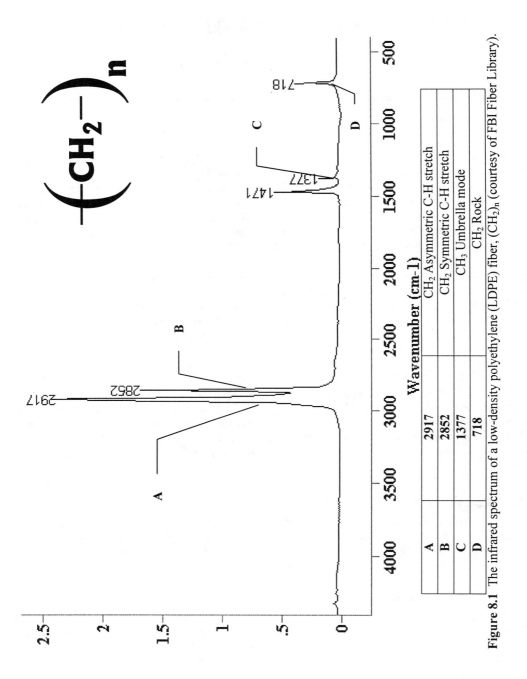

Figure 8.1 The infrared spectrum of a low-density polyethylene (LDPE) fiber, $(CH_2)_n$ (courtesy of FBI Fiber Library).

179

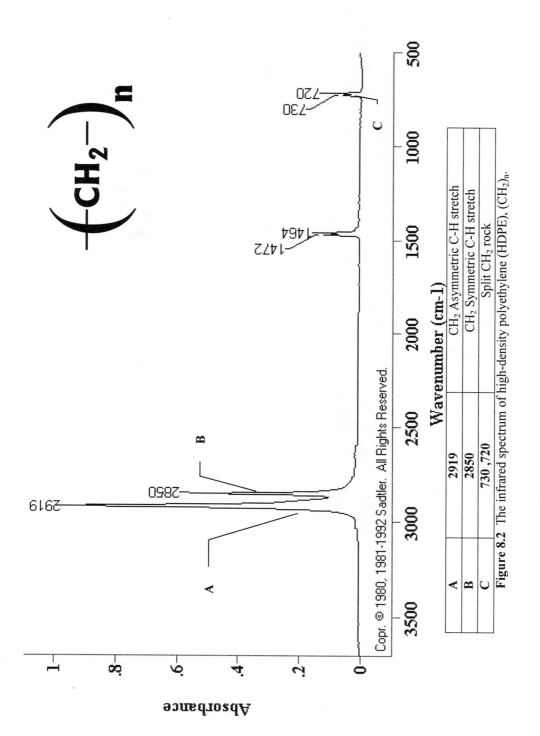

A	2919	CH$_2$ Asymmetric C-H stretch
B	2850	CH$_2$ Symmetric C-H stretch
C	730, 720	Split CH$_2$ rock

Figure 8.2 The infrared spectrum of high-density polyethylene (HDPE), (CH$_2$)$_n$.

Since polyethylene contains only methylene groups, the only infrared bands it displays are CH_2 stretching and bending vibrations. In Figure 8.1, the CH_2 asymmetric and symmetric stretches appear at 2917 and 2852 cm^{-1}, respectively. The methylene scissoring vibration appears at 1468 cm^{-1}, and the rocking vibration appears at 718 cm^{-1}. There are no CH_3 groups in polyethylene, hence no methyl vibrations should be observed. The only way a CH2 chain can be terminated without a methyl group is either to be cyclic or to contain an "infinitely long" chain of methylenes. Polyethylene is an example of the latter. The spectra of pure cyclic alkanes such as cyclohexane have some similarity to the spectrum of polyethylene. However, the cyclic alkanes are generally liquids, whereas polyethylene is a solid. Thus, the physical state of these samples can be used to distinguish between their spectra.

Two different types of polyethylene are commonly manufactured. Low density polyethylene, or LDPE, contains a number of short hydrocarbon side chains terminated with methyl groups. These side groups prevent the methylene chains from crystallizing, giving rise to an amorphous, low-density material. Figure 8.1 is an example of an LDPE spectrum. The methyl groups in the side chains in LDPE often have a small umbrella mode near 1375 cm^{-1}. This band is clearly seen in Figure 8.1 at 1377 cm^{-1}.

The second type of polyethylene, known as high-density polyethylene or HDPE, has very few side chains, and so has very few methyl groups. Its chains pack together easily forming crystalline regions. The spectrum of HDPE is seen in Figure 8.2. Note the lack of an umbrella mode peak near 1375 cm^{-1}. Also, both the methylene scissoring band at ~1460 cm^{-1} and the rocking vibration at 720 cm^{-1} are split into two peaks. Recall from the spectrum of Vaseline in Chapter 2 that this splitting is caused by closely packed methylene chains that interact with each other. This interaction gives rise to the bands at 730 and 720 cm^{-1} seen in Figure 8.2. These bands and the absence of an umbrella mode band can be used to distinguish between low and high-density polyethylene based on their infrared spectra.

B. Polypropylene

Polypropylene (PP) is another common polymer found in everything from food bowls to disposable diapers. Polypropylene is one of the six packaging plastics that are commonly recycled. It is made from the polymerization of propylene ($CH_2=CH-CH_3$), and its chemical formula is $[CH_2-CH-CH_3]_n$. The polymeric backbone of polypropylene consists of alternating methylene and methine (C-H) groups, with each repeat unit containing a pendant methyl group. The spectrum of polypropylene is seen in Figure 8.3. The methyl bands in PP are the asymmetric and symmetric C-H stretches at 2956 and 2875 cm^{-1}, and the umbrella mode at 1377 cm^{-1}. It is easy to distinguish between polypropylene and polyethylene by the presence in the PP spectrum of strong methyl group bands. The CH_2 stretching vibrations in the spectrum of polypropylene appear at 2921 and 2840 cm^{-1}. There is no CH_2 rocking band at 720 cm^{-1} in the spectrum of polypropylene because a methine group separates each CH_2 unit from the one next to it. Although the methine group is present in every repeat unit of PP, it is difficult to determine this from the infrared spectrum of the material.

C. Polystyrene

Polystyrene is another economically important polymer and, is one of the six commonly recycled packaging plastics. Its backbone is the same as polypropylene, consisting of alternating methylene and methine groups. However, each repeat unit in polystyrene contains a pendant benzene ring. The structure and spectrum of polystyrene are seen in Figure 8.4.

The methylene asymmetric and symmetric stretches are seen at 2923 and 2850 cm^{-1}. There is a group of aromatic C-H stretches around 3050 cm^{-1} and benzene ring modes are found at 1600 and 1492 cm^{-1}. The out-of-plane C-H bend of the aromatic ring hydrogens is seen at 756 cm^{-1} and the ring-bending vibration appears at 698 cm^{-1}. These last two bands confirm that polystyrene contains a monosubstituted benzene ring. There is no methylene rocking vibration at 720 cm^{-1} because the methylenes in this molecule are not in a row.

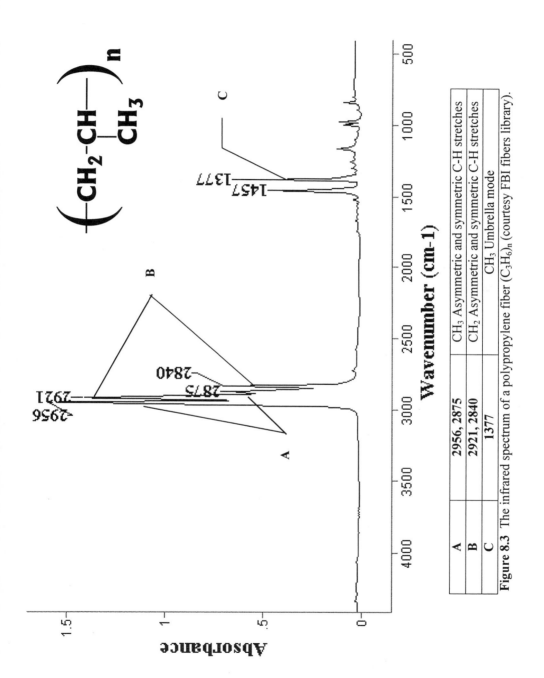

Figure 8.3 The infrared spectrum of a polypropylene fiber $(C_3H_6)_n$ (courtesy FBI fibers library).

A	2956, 2875	CH_3 Asymmetric and symmetric C-H stretches
B	2921, 2840	CH_2 Asymmetric and symmetric C-H stretches
C	1377	CH_3 Umbrella mode

182   Infrared Spectra of Polymers

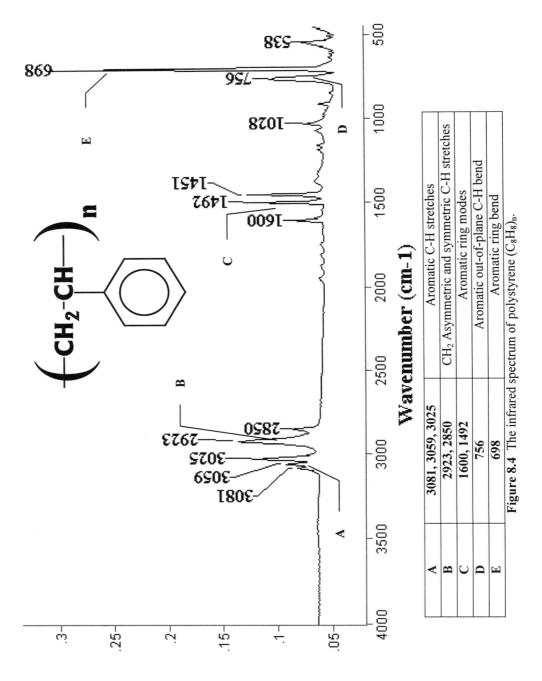

Figure 8.4 The infrared spectrum of polystyrene $(C_8H_8)_n$.

A	3081, 3059, 3025	Aromatic C-H stretches
B	2923, 2850	CH_2 Asymmetric and symmetric C-H stretches
C	1600, 1492	Aromatic ring modes
D	756	Aromatic out-of-plane C-H bend
E	698	Aromatic ring bend

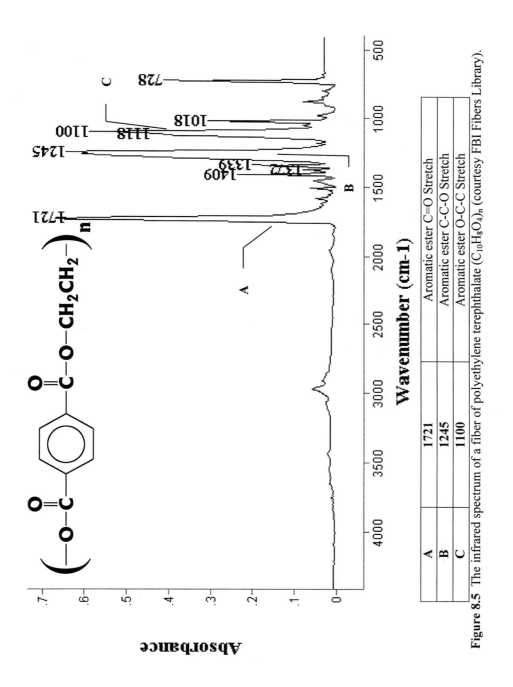

Figure 8.5 The infrared spectrum of a fiber of polyethylene terephthalate $(C_{10}H_8O_4)_n$ (courtesy FBI Fibers Library).

A	1721	Aromatic ester C=O Stretch
B	1245	Aromatic ester C-C-O Stretch
C	1100	Aromatic ester O-C-C Stretch

D. Polyethylene Terephthalate (PET)

By definition, polyesters contain an ester group as part of the polymer backbone. One of the most common polyesters is polyethylene terephthalate (PET), whose structure and spectrum are seen in Figure 8.5. PET is made in quantities of billions of pounds per year, and is one of the six packaging plastics that are commonly recycled. Everything from clothing to soda bottles is made out of this versatile material. PET's tradenames include Dacron® and Mylar®. Like small molecule esters, polyesters follow the rule of three. This means there should be three intense bands in polyester spectra near 1750, 1200, and 1100 cm^{-1}. PET has intense bands at 1721, 1245, and 1100 cm^{-1}. These bands are due to C=O, C-C-O, and O-C-C stretching, respectively. Since PET is an aromatic ester, it follows the rule of three for aromatic esters.

III. Engineering Plastics

These materials are typically more durable and expensive than the recyclable plastics. They are used in a variety of applications including clothing, rope, and as structural members such as windows, car bumpers, and sailboat hulls. The compounds discussed here include nylons, acrylates, polyurethanes, polycarbonates, polyimides, and Teflon®.

A. Polyamides (Nylons)

Polyamides comprise a diverse but important family of polymeric materials. Manmade polyamides include the entire family of nylons, as well as polymers such as Kevlar® and Nomex®. These materials give strong, light fibers and are used in clothing, stockings, and rope. Polyamides are ubiquitous in the natural world. Every protein in every living cell in the world is a polyamide. Materials such as skin, hair, insect bodies, silk, and spider webs are all examples of polyamides.

As mentioned in Chapter 5, most polyamides contain secondary amide linkages. In these linkages, the amide nitrogen has one N-H bond and two N-C bonds. The structure and spectrum of a common commercial polyamide, Nylon-6,6, is shown in Figure 8.7. The "-6,6" in the name of this material means there are two methylene chains in this molecule, each of which contains six CH$_2$ units. Nylon-3,5 would contain alternating hydrocarbon chains of 3 and 5 methylene units, respectively. Nylon-6,6 is a secondary amide, so it shows a single N-H stretch at 3301 cm^{-1}, a carbonyl C=O stretch at 1641 cm^{-1}, and an N-H bend at 1541 cm^{-1}. The two intense bands at ~1640 and 1540 cm^{-1} are diagnostic for the secondary amide structural unit. These two bands in the spectrum of a polymeric sample are a strong indication that nylon is present.

B. Acrylates

The members of the acrylate family of polymers are made from acrylic acid and its derivatives. The structure of acrylic acid is seen in Figure 8.6.

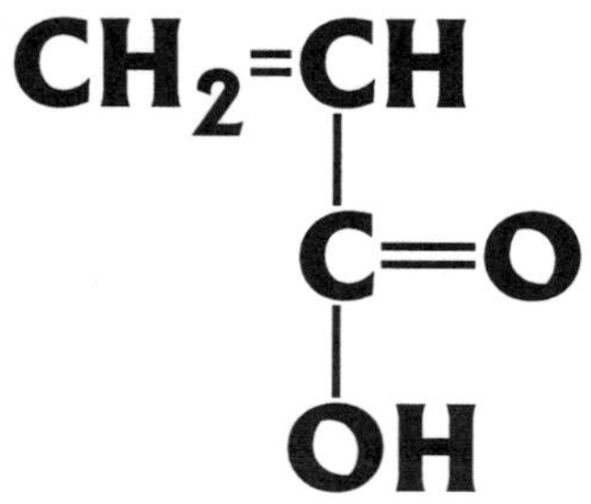

$$CH_2=CH$$
$$|$$
$$C=O$$
$$|$$
$$OH$$

Figure 8.6 The structure of acrylic acid.

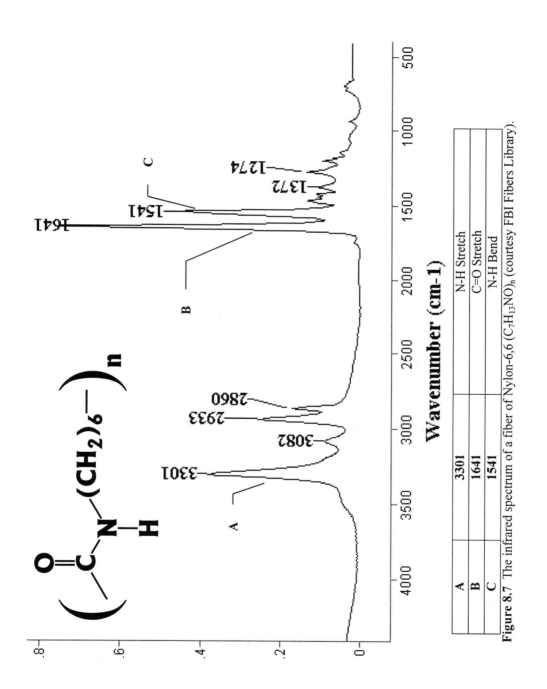

Figure 8.7 The infrared spectrum of a fiber of Nylon-6,6 ($C_7H_{13}NO$)$_n$ (courtesy FBI Fibers Library).

A	3301	N-H Stretch
B	1641	C=O Stretch
C	1541	N-H Bend

185

186 Infrared Spectra of Polymers

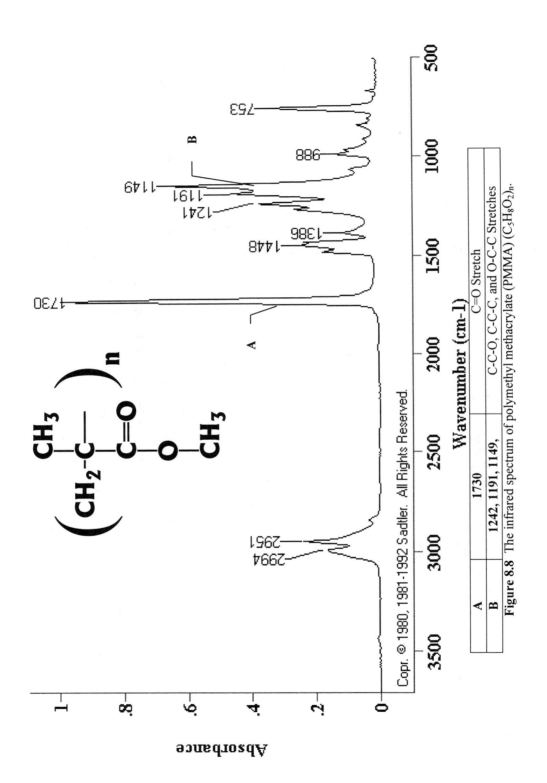

Figure 8.8 The infrared spectrum of polymethyl methacrylate (PMMA) $(C_5H_8O_2)_n$.

A	1730
B	1242, 1191, 1149,

	C=O Stretch
	C-C-O, C-C-C, and O-C-C Stretches

The vinyl group in this molecule is polymerized to give backbones of alternating methylene and methine groups. The nature of the pendant groups distinguishes acrylates from each other.

Polymethyl methacrylate (PMMA) is an important member of the acrylate family. It is also known as Plexiglass® and Lucite®. It is a tough, durable polymer that is transparent to visible light, and is often used in applications where traditional window glass is used. PMMA has a methyl group and an ester group attached to the polymeric backbone. However, PMMA is not considered a polyester because the ester group is not part of the backbone. The infrared spectrum and structure of PMMA are seen in Figure 8.8. The carbonyl stretch of the ester is seen at 1730 cm^{-1}. The clump of bands between 1250 and 1150 cm^{-1} is made up of contributions from C-C-O, O-C-C, and C-C-C stretching vibrations. Other members of the acrylate family include polymethacrylic acid (used in soft contact lenses), polyacrylonitrile, and ABS rubber (an acrylonitrile, butadiene, styrene copolymer).

C. Diisocyanates and Polyurethanes

Diisocyanates are the precursors to the urethane family of polymers. The structural frameworks of diisocyanates and urethanes are shown in Figure 8.9

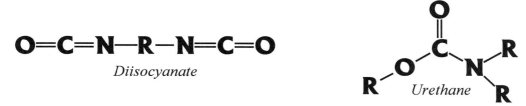

Figure 8.9 The structural frameworks of the diisocyanate and urethane functional groups.

Note that a diisocyanate contains two isocyanate functional groups, R-N=C=O.

Polyurethanes are made by the reaction of a diisocyanate and a diol (R(OH)$_2$). Polyurethanes are made into polyurethane foam and are used in finishes applied to wood. Urethanes contain a carbonyl group with an oxygen atom and a nitrogen atom attached to the carbonyl carbon. The left half of a urethane resembles an ester, and the right half resembles an amide. Like amides, urethanes are denoted primary if one carbon is attached to the nitrogen atom, secondary if two carbon atoms are attached to the nitrogen atom, and tertiary if three carbon atoms are attached to the nitrogen. For a urethane to be part of a polymer chain, it must have at least two carbon atoms attached to the nitrogen. Thus, all polyurethanes are either secondary or tertiary.

The infrared spectrum of a diisocyanate is shown in Figure 8.10. The most important group wavenumber for isocyanates is the asymmetric stretching of the N=C=O bonds. This band is very intense, and appears between 2280 and 2240 cm^{-1}. There is no mistaking this band in Figure 8.10 at 2277 cm^{-1}. The position of the N=C=O stretch is usually unaffected by conjugation.

The carbonyl (C=O) stretches of secondary and tertiary urethanes show up from 1725 to 1705 cm^{-1}, and from 1690 to 1680 cm^{-1}, respectively. Like a secondary amide, a urethane has a high wavenumber N-H bending vibration, which typically appears between 1540 and 1520 cm^{-1}. Like an ester, there is a urethane C-O single bond stretch around 1250 cm^{-1}. A secondary urethane contains a single N-H bond, which gives rise to an N-H stretching vibration from 3340 to 3250 cm^{-1}.

Figure 8.11 is the infrared spectrum of a polyurethane/polyether foam rubber. Only the bands due to the polyurethane have been marked. The material is a secondary urethane because of the N-H stretch at 3326 cm^{-1}. The carbonyl stretch is at 1705 cm^{-1}, the N-H bend at 1531 cm^{-1}, and the C-O stretch is at 1222 cm^{-1}. Note that a secondary urethane has a pair of strong bands at ~1700 and ~1550 cm^{-1} like a secondary amide. In both cases these bands are due to a C=O stretch and N-H bend.

188 Infrared Spectra of Polymers

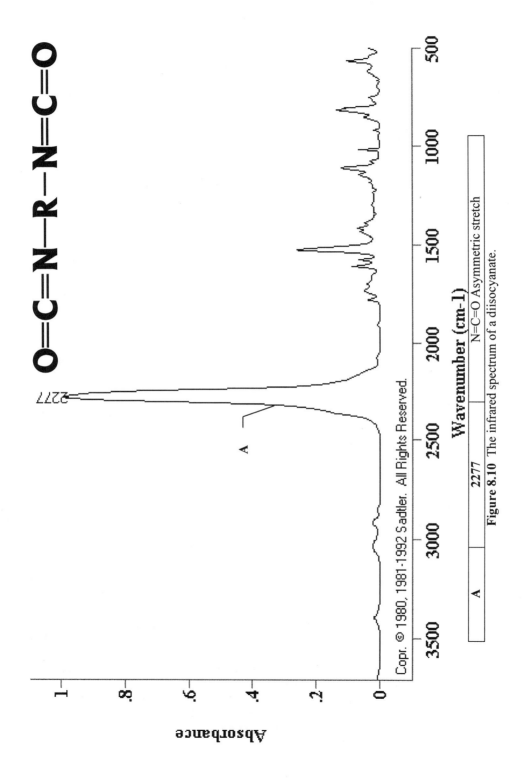

Figure 8.10 The infrared spectrum of a diisocyanate.

| A | 2277 | N=C=O Asymmetric stretch |

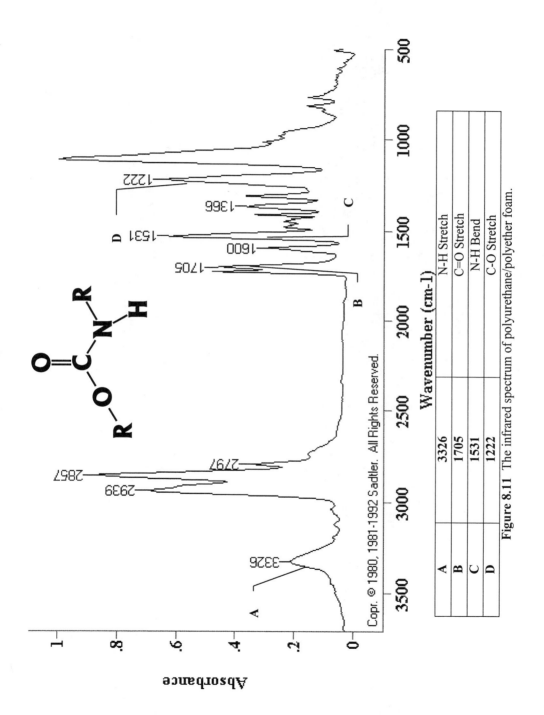

Figure 8.11 The infrared spectrum of polyurethane/polyether foam.

A	3326	N-H Stretch
B	1705	C=O Stretch
C	1531	N-H Bend
D	1222	C-O Stretch

Also, note that secondary urethanes have three strong bands at ~1700, 1550, and 1250 cm^{-1}, a "rule of three" whose pattern is similar to the three intense bands found in ester spectra. Table 8.1 lists the group wavenumbers for the isocyanate and urethane functional groups.

Table 8.1 Group Wavenumbers for Diisocyanates and Urethanes

Vibration	Wavenumber (cm^{-1})
Isocyanate asymmetric N=C=O stretch	2280-2240
Secondary Urethane N-H Stretch	3340-3250
Secondary Urethane C=O	1725-1705
Tertiary Urethane C=O	1690-1680
Secondary Urethane N-H Bend	1540-1520
Urethane C-O Stretch	~1250

D. Polycarbonates

Polycarbonates are important engineering plastics because they are tough and durable. The most common polycarbonate is sold under the tradename "Lexan®." Its chemical structure and infrared spectrum are seen in Figure 8.12. The characteristic carbonyl stretch of a diaromatic carbonate is seen at 1777 cm^{-1}, and the C-O stretch is at 1230 cm^{-1}.

E. Polyimides

Polyimides are tough engineering plastics increasingly used as coatings for glass capillary tubes, wire, and fiber optic cables. As discussed in Chapter 5, the two carbonyl groups in a cyclic imide give rise to two C=O stretching bands between 1800 and 1700 cm^{-1}. These vibrations correspond to the carbonyl groups stretching in phase and out of phase with each other. The higher wavenumber band is less intense in the spectra of cyclic imides. The infrared spectrum of Kapton®, a cyclic polyimide, is seen in Figure 8.13. Note that Kapton also contains an ether linkage, but only the bands due to the imide are labeled. The double carbonyl stretch at 1777 and 1726 cm^{-1} is diagnostic for cyclic imides. There are no N-H bonds in this imide, and the C-N stretch is probably too weak to be seen.

F. Polytetrafluoroethylene (Teflon®)

Polytetrafluoroethylene is better known by its tradename, Teflon®. Teflon is famous for its durability and nonstick properties. The structure of Teflon is $[CF_2]_n$, and it consists solely of C-F and C-C bonds. The strength of these bonds gives Teflon its unique properties. The structure and infrared spectrum of polytetrafluoroethylene are seen in Figure 8.14. As with other halogenated compounds, it has intense bands at low wavenumber. The bands at 1212 and 1155 cm^{-1} in Figure 8.14 are due to the stretching of C-F bonds. The C-F bend would normally be seen around 650 cm^{-1}, but the detector used to obtain this spectrum cut off at 750 cm^{-1}.

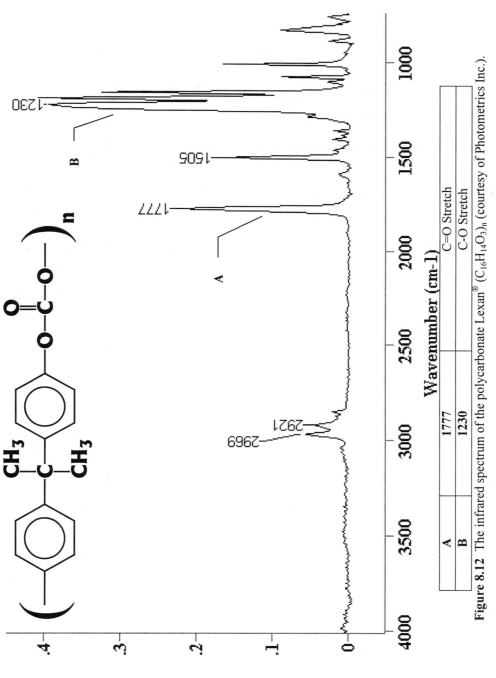

Figure 8.12 The infrared spectrum of the polycarbonate Lexan® $(C_{16}H_{14}O_3)_n$ (courtesy of Photometrics Inc.).

A	1777	C=O Stretch
B	1230	C-O Stretch

192 Infrared Spectra of Polymers

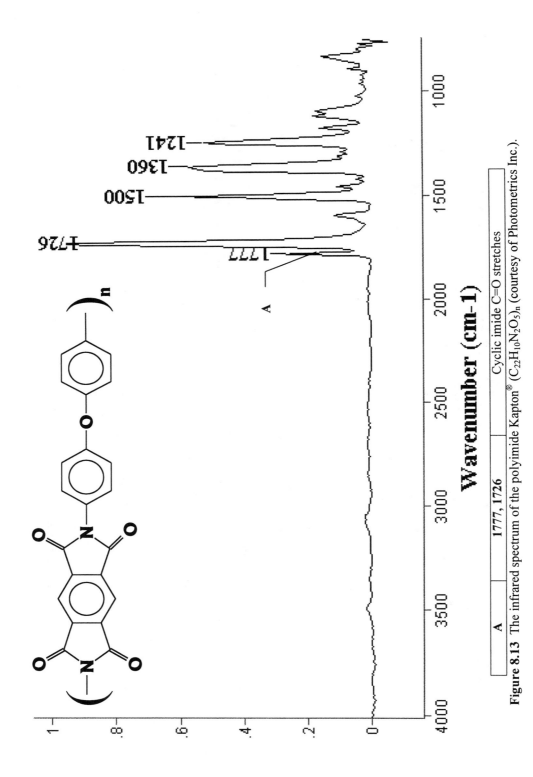

Figure 8.13 The infrared spectrum of the polyimide Kapton® $(C_{22}H_{10}N_2O_5)_n$ (courtesy of Photometrics Inc.).

| A | 1777, 1726 | Cyclic imide C=O stretches |

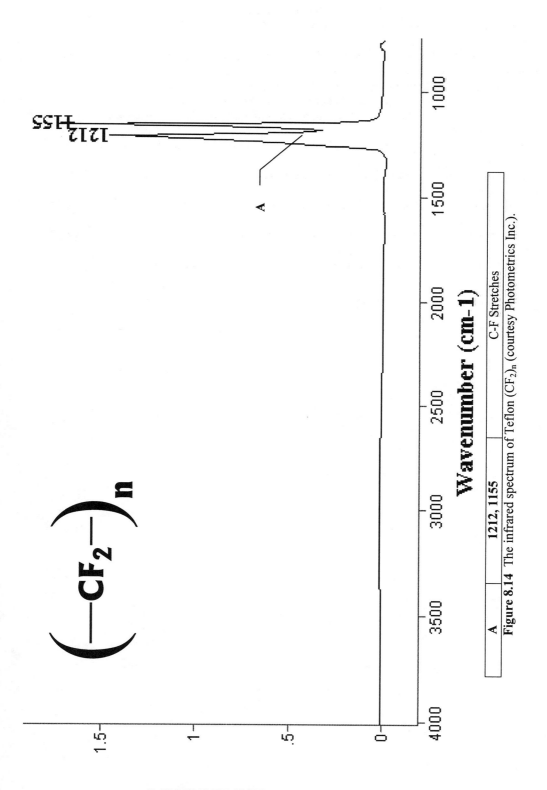

Figure 8.14 The infrared spectrum of Teflon $(CF_2)_n$ (courtesy Photometrics Inc.).

| A | 1212, 1155 | C-F Stretches |

Bibliography

L.J. Bellamy, *The Infrared Spectra of Complex Molecules, Third Edition*, Wiley, New York, 1975.

R. Silverstein, G. Bassler, T. Morrill, *Spectrometric Identification of Organic Compounds, Fourth Edition*, Wiley, New York, 1981.

D. Lin-Vien, N. Colthup, W. Fately, J. Grasselli, *Infrared and Raman Characteristic Frequencies of Organic Molecules*, Academic Press, Boston, 1991.

N. Colthup, L. Daly, S. Wiberley, *Introduction to Infrared and Raman Spectroscopy*, Academic Press, New York, 1990.

F. Miller, J. Graselli, *Lecture Notes for Bowdoin College Infrared Short Course*, Brunswick, Maine, 1992.

B.C. Smith, *Fundamentals of Fourier Transform Infrared Spectroscopy*, CRC Press, Boca Raton, 1996.

H. Mantsch, D. Chapman, Eds., *Infrared Spectroscopy of Biomolecules*, Wiley, New York, 1996.

A. Streitweiser, C. Heathcock, *Introduction to Organic Chemistry*, Macmillan, New York, 1976.

G. Odian, *Principles of Polymerization*, McGraw-Hill, New York, 1970.

Chapter 9

Spectral Interpretation Aids

The first eight chapters of this book focused on helping readers obtain the skills and knowledge needed to interpret infrared spectra on their own. As you have probably discovered by this point, interpreting spectra can be difficult. Problems that can be encountered include molecules not following "the rules," bands from different functional groups overlapping, and noisy or poorly resolved spectra. Fortunately, help is available in the form of interpretation aids to make your job easier. This chapter will describe some of these aids, and has been placed at the end of the book because interpretation aids should be used <u>after</u> you have had time to study a spectrum. These aids are not a substitute for knowledge and experience. On the contrary, the more you know about interpreting spectra, the more information can be gleaned by using interpretation aids. The aids that will be discussed include collections of spectra published in atlases, spectral subtraction, library searching, software programs, and the Internet.

I. Spectral Atlases

Infrared spectral atlases are collections of infrared spectra published in book form. Typically, the spectra are organized by functional group. For example, there may be sections on hydrocarbons, esters, and amines. The most common sources of atlases are companies that also sell spectral libraries (see below). The utility of atlases is their organization; by compiling all the spectra of related compounds in one place, it is easy to see at a glance what peaks are diagnostic for a family of molecules. There are only ~50 spectra shown in this book covering a variety of chemical structures; an atlas can provide hundreds of spectra of a specific functional group. Once familiar with the group wavenumbers of a functional group, seeing hundreds of examples can give one a feel for the patterns of bands in spectra.

Atlases can also help in identifying unknowns. For example, you may suspect that a sample contains an aromatic amine. By comparing the spectrum of your sample to that of a collection of aromatic amines, you can quickly discern the similarities or differences between the spectra. Another type of problem that atlases are useful in solving is identities. As mentioned in Chapter 1, taking a spectrum of a material and comparing it to a reference spectrum of a known material to confirm an identity is one of the most common uses of infrared spectroscopy. Atlases can serve as a source of reference spectra for identities. For example, if you want to know if a sample is acetone or not, simply take its spectrum and compare it to the acetone spectrum contained in an atlas.

The old-fashioned way of comparing spectra (before computers) was to make a photocopy of the atlas spectrum, overlay it with the spectrum in question, and then hold the two up to the light to see them both clearly. Products known as light boxes are still sold for this purpose, allowing one to place two spectra on top of each other, and illuminate them from below to see subtle differences between spectra. Now, if the two spectra you need to compare are stored in digital format in a computer, comparisons can be made by plotting the spectra on the same piece of paper, or overlaying the spectra on the computer screen. However, this old-fashioned spectroscopist still finds it helpful on occasion to plot out spectra and literally hold them up to the light for comparison purposes. Do not dismiss the power and simplicity of this technique for performing identities.

II. Spectral Subtraction

Spectral subtraction is one of the main tools used to attack the problem of mixture spectra. One way of simplifying the spectrum of a complex mixture is to simplify the chemical composition of the mixture through purification. If this is not an option, spectral subtraction allows one to simplify the spectrum itself. This is done by taking the spectrum of what is typically a pure material, subtracting it from the mixture spectrum, and thereby removing the bands due to this material. The technique can be very useful for removing solvent bands, carbon dioxide and water vapor bands, or any bands due to an unwanted component.

When performing a subtraction, the spectrum of the mixture is called the *sample spectrum*, and the spectrum of a component in the mixture to be subtracted from the sample spectrum is called the *reference spectrum* (sometimes called the subtrahend). Spectra to be subtracted must be in units that are linearly proportional to concentration, which typically means absorbance units. Transmission and single beam spectra should not be subtracted, since their peak heights and areas are not linearly proportional to concentration. In addition, spectral subtraction assumes the validity of Beer's law for the spectra involved.

The principle behind spectral subtraction is simple: the absorbance values of the reference are subtracted point by point from the absorbance values of the sample. For example, if the absorbance of the sample at 3400 cm^{-1} is 0.7, and the absorbance of the reference at the same wavenumber is 0.4, the subtraction result at 3400 cm^{-1} is 0.3 absorbance units. The subtraction result is a plot of the difference in absorbance between the two spectra versus wavenumber.

Performing subtractions in such a direct manner ignores the fact that the concentration of the reference material may have been different in the sample and reference spectra. For example, the concentration of water in its pure form is higher than in a soap and water solution. This is because the water is actually diluted when the soap is dissolved in it. To perform the subtraction without taking into account the concentration differences of water in the two spectra would mean that there would be water bands left behind in the result. To get around this problem, the reference spectrum is multiplied times a *subtraction factor*. Adjustment of the subtraction factor allows the user to scale the absorbances in the reference spectrum and minimize features due to the reference spectrum component. The general equation used to perform subtractions is as follows

$$(Sample) - (Sub\ Factor) * (Reference) = Result$$

The absorbance values of the reference spectrum are multiplied by the subtraction factor (also called the scale factor), then subtracted point by point from the absorbance values of the sample spectrum. Best results are obtained when the absorbances of the sample and the reference are about the same, and if a subtraction factor close to one is used.

A. Optimizing the Subtraction Factor

Before performing a subtraction, visually compare the sample and reference spectra. Bands common to both spectra are due to the reference material, while bands that appear only in the sample are of interest and should be preserved in the subtraction result. The subtraction factor is the only user adjustable parameter in the subtraction process. It is not always obvious what value of the subtraction factor will give the best results. To set the subtraction factor properly

1. Choose a spectral feature or features common to both the sample and reference spectra. Make sure that these features are less than 0.8 absorbance units. Bands more intense than this may not subtract properly, while bands at this intensity or lower usually do subtract properly.

2. Adjust the subtraction factor interactively until the feature or features chosen in step 1 appear flat and become part of the baseline.

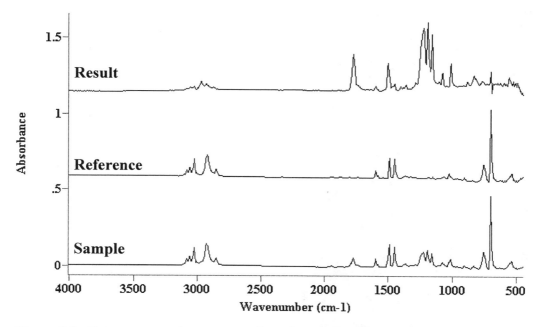

Figure 9.1 Bottom: a sample spectrum of a mixture of polystyrene and a polycarbonate. Middle: a reference spectrum of pure polystyrene. Top: a subtraction result using a subtraction factor of 1.75.

If there are several absorbance bands that meet the criteria stated in step 1, you can try using each band in turn to adjust the subtraction factor. It may turn out that using one band may give better results than another. Alternatively, a subtraction factor set by using several bands at a time may give better results than using just one band. There is some trial and error involved in choosing a feature to minimize and in adjusting the subtraction factor.

The process of subtraction is illustrated in Figure 9.1. The bottom spectrum is of a polystyrene/polycarbonate mixture, the middle spectrum is of pure polystyrene, and the top spectrum is of the subtraction result. There are bands in the sample spectrum around 1700 and 1200 cm^{-1} that are from the polycarbonate, but there is not enough information present to determine the structure of the polycarbonate. The feature in the polystyrene spectrum at 1029 cm^{-1} was chosen because it is common to both spectra and is less than 0.8 absorbance units high. The subtraction factor was adjusted to a value of 1.75, at which point the peak at 1029 cm^{-1} was no longer visible. The result shows many bands between 2000 and 800 cm^{-1} that are definitely due to the polycarbonate and would be very useful in determining its molecular structure.

Figure 9.2 is an expansion of the spectra seen in Figure 9.1. Note that by comparing the sample and reference spectra, polystyrene peaks at 1028 and 1068 cm^{-1} can be identified, as well as polycarbonate bands at 1080 and 1015 cm^{-1}. These bands are partially overlapped. In the subtraction result in the top of Figure 9.2, the polystyrene bands are gone, but the polycarbonate bands at 1081 and 1015 cm^{-1} are clearly visible. This result illustrates the ability of spectral subtraction to work well even on spectra with overlapped bands.

B. Subtraction Artifacts

Ideally, a subtraction result should contain no bands from the reference spectrum, be free of artifacts, and have a flat baseline. In reality, there are two types of artifacts that may appear in subtracted spectra. Figure 9.3 is an expansion of the spectra seen in Figure 9.1, and illustrates both types of subtraction artifact.

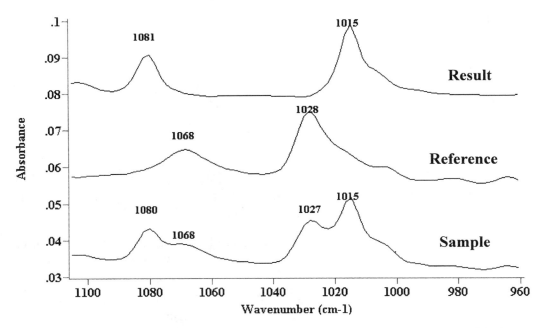

Figure 9.2 Expansion of the spectra seen in Figure 9.1. Bottom: mixture spectrum of polystyrene and a polycarbonate. Middle: reference spectrum of pure polystyrene. Top: subtraction result. Note that even overlapped bands can subtract out properly.

The first artifact that can appear in subtraction results are unsubtracted reference bands. These can be identified by closely comparing the result spectrum to the reference spectrum. Any features common to these two spectra are unsubtracted reference bands. An example of this type of subtraction artifact is the feature at 700 cm^{-1} seen at the top of Figure 9.3. This band did not subtract out completely, and so is seen in the result spectrum. This feature corresponds to the most intense band in the original polystyrene spectrum, and did not subtract out properly because the height of this band does not vary linearly with concentration. This feature illustrates it is often the most intense bands in the reference spectrum that do not subtract out properly. Unsubtracted reference bands may point up or down, but no amount of manipulation of the subtraction factor can eliminate their presence. Reducing the concentration or pathlength of the sample may bring intense absorbances into a range where they may subtract out properly. However, the usual way of dealing with these artifacts is to learn to recognize and ignore them. Use artifact-free regions of the result spectrum for spectral interpretation.

The second type of subtraction artifacts are derivative shaped peaks. These peaks point above and below the baseline, and are similar in appearance to the mathematical derivative of an absorbance band. The shape of the feature at 700 cm^{-1} at the top of Figure 9.3 illustrates the shape of these artifacts. Derivative shaped bands are caused by wavenumber or bandshape shifts in the spectrum of the reference material compared to the spectrum of a mixture. In the example in Figure 9.3, the intense polystyrene band has a slightly different bandshape in the pure form than in the mixture with the polycarbonate. The two bands are not perfect overlays, so will not subtract properly. The part of the derivative band that is positive is the wavenumber range where the sample spectrum absorbance is more intense than the reference spectrum. The negative part of the derivative band is the wavenumber range where the sample spectrum absorbance is less intense than the reference spectrum.

The band position and shape shifts that cause derivative bands are due to chemical interactions in a sample. The chemical interactions of polystyrene in its pure form are slightly different from when it is mixed with a polycarbonate. The slightly different chemical environments in these two samples cause a bandshape shift, giving rise to the artifact seen in Figure 9.3. Since nothing can be done to prevent these chemical interactions, nothing can be done about derivative shaped bands except to recognize and ignore them.

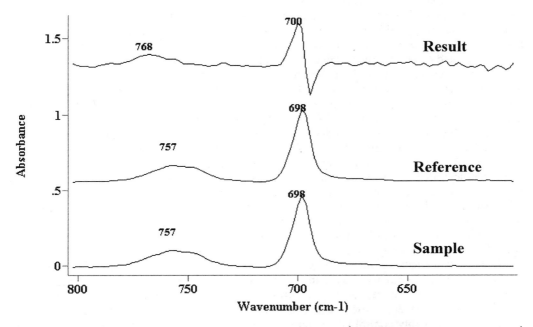

Figure 9.3 The spectra in Figure 9.1 expanded in the 700 cm^{-1} range. The feature at 700 cm^{-1} in the result spectrum is an example of both types of subtraction artifact. It is an unsubtracted reference band and has a derivative shape.

Very few subtractions are perfect, and the artifacts discussed here are common. However, by following the process of adjusting the subtraction factor outlined above, and by being aware of the existence of artifacts that can be generated as a result of performing a subtraction, this form of spectral manipulation should yield useful information for you.

III. Spectral Library Searching

Spectral library searching is a widely used interpretation aid. The idea behind library searching is to mathematically compare an unknown (or sample) spectrum against a collection of known spectra contained in electronic form in a library. The result of this comparison is a number called the "hit quality index" (HQI), which is a direct measure of how similar two spectra are to each other. In the jargon of library searching a good hit occurs when the match between two spectra is close, and a bad hit occurs when two spectra are very dissimilar. If a good hit is found, it can be assumed that the unknown sample and known sample have similar compositions.

Before we discuss the attributes of library searching, it is important to realize what the results of a library search mean. The computer will always find a match for your unknown among the spectra searched, even if it is a lousy match. The existence of a match by itself is meaningless. It is up to you, the user, to interpret the results of a library search. The author has seen too many instances of people searching a library, finding a match, then declaring the sample has been identified without bothering to look closely at the search results. You should ALWAYS visually compare the unknown spectrum to the library spectra matched to your sample. Often the computer may indicate two spectra are well matched, yet you may be able to see vast spectral differences between them. Other times, the computer may say a match is poor, but you may be able to see enough similarity between the spectra to help in identifying what is in a sample. In all cases, trust your own judgement and use your own knowledge of the sample and of spectroscopy to guide you in interpreting search results. Library searching is not a substitute for good science or interpretive skills; it is a tool to enhance the skills of the analyst.

This discussion of library searching will be limited to what is known as a full spectrum search. This method uses the entire sample spectrum, and compares it to each library spectrum in its entirety. A typical infrared spectrum contains thousands of data points and using all of them in a library search greatly increases the odds of obtaining an accurate search result. In essence, if the data points are available, they should be used in the library search.

The first question that needs to be answered about library searching is where to obtain spectral libraries. In the past, spectral libraries were available only in book form. Today, spectral libraries come in digitized form on floppy or compact disks and are copied to your computer's hard disk. Many FTIR companies sell spectral libraries. In addition, two companies, Sadtler Division of Bio-Rad (Philadelphia, PA) and Aldrich Chemical Company (Milwaukee, WI), sell digitized infrared spectral libraries. There are literally hundreds of thousands of spectra commercially available. Libraries are available organized by sample type. For example, polymer, gas phase, inorganic, organic, and adhesives libraries are available. It is best to match the samples in your library to the type of samples you most frequently analyze.

Despite the availability of commercial infrared libraries, one of the best sources of libraries is you, the user of an infrared spectrometer. Making your own libraries is a good idea because only you have access to the kinds of samples that are typical in your work. For example, if you identify an unknown today and put it in a library, you will always have that information at your disposal. If you encounter the same sample 3 years from now, you will be able to quickly and easily identify it since it will show up as a good hit when the unknown is searched. When making your own libraries, spectra of samples that are related to each other are best kept together. For instance, a user generated library could contain the spectra of all the raw materials that a quality control lab analyzes, spectra of all the batches of a finished product, or all the polymers ever synthesized by an organic chemistry group. All the spectra in a library must be of the same resolution and cover the same wavenumber range. Library spectra act as de facto references, so you should be certain of the identity of samples whose spectra you place in a library. Additionally, library spectra should have a high SNR, be free of artifacts, and be in absorbance units.

Spectra added to a library are usually baseline corrected. Baseline slope is usually caused by instrument or sampling problems, and it is undesirable to have these contribute to search results. Library and unknown spectra are also often *normalized*. Normalization means all the absorbances in a spectrum are divided by the largest absorbance in the spectrum. For example, if the biggest peak in a spectrum has an absorbance of 0.8, all the absorbances in the spectrum are divided by 0.8. The intensity scale for all normalized spectra is from 0 to 1. The purpose of normalization is to remove differences in peak heights between spectra acquired under different conditions (i.e., different concentrations and pathlengths). Essentially, it allows an apples-to-apples comparison to be made.

It is important to select the proper libraries to search. Unknown spectra should be searched against libraries that contain spectra of samples similar to the unknown. Essentially, you want to search likes against likes. For instance, if you search a gas phase spectrum against a polymer library, you will not get good results because polymers do not exist in the gas phase! However, you may want to search a polymer spectrum against a polymer library and a general organics library, since many polymers are organic materials.

A. Search Algorithms

Before unknown and library spectra are compared to each other, the unknown spectrum is usually normalized and baseline corrected. Next, the unknown spectrum is compared to each library spectrum using a *search algorithm* to generate the hit quality index. Search algorithms are different ways of mathematically comparing two spectra to each other. There are many search algorithms available, depending on the software used to perform a search.

Here is how one simple search algorithm works, the absolute value algorithm. First, a library spectrum is subtracted from the sample spectrum. The result of this calculation is called a residual. The size of the residual is directly related to how similar two spectra are to each other. Identical spectra will have a residual of zero (a straight line); dissimilar spectra will have residuals with large features in them. The size of the residual is calculated by taking the absolute value of each data point in the residual, adding them together, and then dividing by the total number of data points (taking the average). This calculation gives the HQI. Other search algorithms that are commonly used have names like correlation, Euclidean distance, first derivative, and least squares. Their mathematics are more complicated than the absolute value algorithm, but they all produce hit quality indices. It is not always obvious which search algorithm will work best for a particular sample. It is common to begin with the correlation or Euclidean distance algorithms. If these fail to produce good results, it is worth trying other algorithms to see if the search results can be improved.

B. Interpreting Search Results

The ultimate product of a search is a listing of the best matches called a *search report*. The search report is organized by HQI, and often contains the name, index number, library name, HQI, and other important information about a library spectrum. Figure 9.4 is an example of a report from the search of a spectrum of pure chloroform. The correlation search algorithm was used.

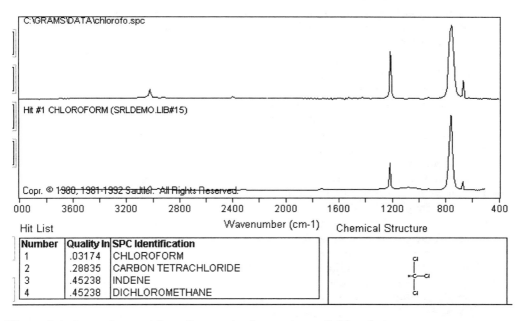

Figure 9.4 A search report from the search of a spectrum of chloroform.

Note that the two spectra look almost identical, so this is a good search result. However, it must be noted that you will obtain a good match only if a spectrum similar to that of your sample happens to be in one of the libraries searched. This is why it is so important to search likes against likes. If there are no good matches in a search report, library searching will disclose very little about a sample. Note that the HQI for the best hit in Figure 9.4 is 0.032. In this particular software, 0.0 is a perfect hit and 1414 is the worst possible hit. A perfect match is rare, and usually occurs only if a copy of a spectrum already in a library is searched against that library.

The value of the HQI is of utmost importance. One of the problems with HQIs is making sense out of the many HQI systems used by different software vendors. Depending on your software package, a value of 0.0, 1.0, 10, 100, or 1000 may indicate a perfect hit. The only way to make sense of the hit quality indices generated by a specific software package is to precisely know what values correspond to a best and worst hit. Only then can you interpret the search report. For example, in a system where 1000 is a perfect match, it is easy to determine that an HQI of 800 is a good match, and an HQI of 200 is a lousy match.

It is straightforward to calculate "percent similarities" using the HQIs in any software package. In the previous example, an HQI of 800 out of 1000 was called a good hit. The percent similarity here is simply 800/1000 or 80%. In general, a percent similarity of 80% or greater is considered good, and means there is enough similarity between the spectra to give useful information about the composition of an unknown. If the percent similarity is 50% to 80%, the search results are considered marginal. Visual comparison may yield suggestions as to the types of molecules or functional groups present in a sample, but sample identification is impossible with this low quality of search results. If the percent similarity is less than 50%, the search results are considered poor, and there is probably little useful information to be gained from the results. Use these percent similarity measurements, along with visual comparison of sample and library spectra, to help you make sense out of a search report.

In addition to the absolute value of the HQI, the relative difference between HQIs in a search report is important. A bunch of hits clustered around one number contains different information than obtaining one good hit and a bunch of bad hits. The percent difference between HQIs is calculated by taking the difference in HQI for two consecutive hits, and dividing the difference by the smaller HQI. For example, in the search report in Figure 9.4, the relative difference in HQI for hits 1 and 2 is (0.031 - 0.29)/(0.031) or 835%. This means that the second hit is significantly different from the first hit, and that the first match should be the only one taken seriously.

C. Subtract and Search Again

In addition to determining the identity of a pure unknown, library searching has some utility in identifying components in mixtures. When searching the spectrum of a mixture, if one of the major components in the unknown turns up as a library hit, you can subtract the component's spectrum from the unknown and search again. With luck, the spectrum of a second component will be a top hit, which means the identity of the second component can be obtained. This is known as subtract and search again. This process is illustrated in Figures 9.5 and 9.6.

The spectrum of a mixture of benzene and cyclohexane was searched giving the search results seen in Figure 9.5. The correlation algorithm was used. Note that cyclohexane is the best hit, although the second hit is within 20% of the first. Visual examination of the spectra, in addition to the fact that the second hit is a solid and the sample is a liquid, would qualify cyclohexane as the best match. The library spectrum of cyclohexane was subtracted from the mixture spectrum and the subtraction result was searched. The results of this search are seen in Figure 9.6. The correlation algorithm was used. Note that benzene is the best match, hundreds of times better than the next closest hit. The library searched in this example contained many other aromatic compounds, so it was not obvious beforehand whether or not this example would give such good results. The subtract and search again technique may not be this simple with complex mixture spectra. From a practical viewpoint, this process can identify only two or three components in a mixture. However, it is another useful tool in the struggle to make sense out of mixture spectra.

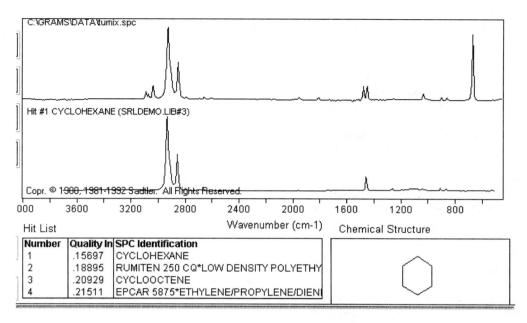

Figure 9.5 The search results obtained with the spectrum of cyclohexane and benzene. The correlation algorithm was used.

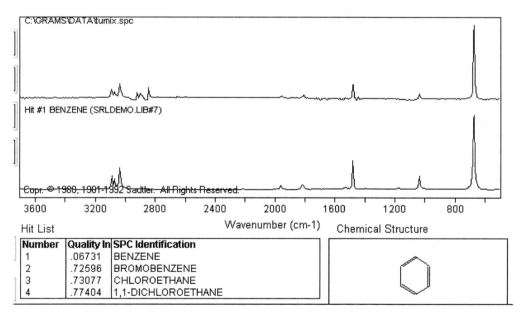

Figure 9.6 The search results obtained after subtracting a cyclohexane spectrum from the spectrum of a cyclohexane/benzene mixture.

IV. Infrared Interpretation Software Packages

There are a number of software packages available today to help teach the art and science of spectral interpretation or that will actually help you interpret a spectrum. The vendors for these packages include most of the major FTIR instrument manufacturers, as well as the better known scientific software companies. These software packages can be divided into two groups.

The first are what will be called interpretation tutorials. These software packages typically come with a collection of spectra, molecular diagrams, and animations of vibrations. These packages allow one to look at the spectrum of a molecule and its structure on the same screen. By clicking on a band in the spectrum, an animation of the vibration that gave rise to that absorbance is displayed. The spectra and vibrations of different functional groups can be observed by looking at different molecules.

These tutorial programs are an excellent teaching tool and complement the information contained in this book. Their strong point is the immediate connection they make between bands and vibrations, making it easier to remember what functional groups absorb where. The limitation of these tutorial programs is the limited number of molecules for which spectra, diagrams, and animations are available. Some molecules and types of functional groups will, by force of necessity, not be covered in these software programs. Additionally, when faced with an unknown spectrum, the small collection of spectra in these programs will make the chances of finding a similar spectrum small. So, these programs are an excellent introduction to the art and science of interpretation, but do not contain enough data to act as a reference source of infrared data.

The second type of interpretation software programs are "expert programs." These programs actually know most of the rules of spectral interpretation taught in this book, and attempt to guide the user in the interpretation of unknown spectra. By importing a spectrum into the program and clicking on a specific peak, the program will display a list of the functional groups most likely associated with that band. The functional groups will be listed in order from highest to lowest probability, like the list of hit quality indices in a search report. Additionally, other places where these functional groups absorb will be shaded, allowing the user to look for other bands that may be present to help identify the functional group in question. These expert programs are truly attempts to "hard wire" the information stored in an experienced spectroscopist's brain into a computer program.

The author has found these expert programs quite useful in figuring out the spectra of unknown samples, and they will be a useful adjunct in any lab where unknown interpretations are regularly performed. Expert programs have not yet evolved to the point where they can replace a real live expert. Additionally, the results of an expert program must be interpreted just like the results of a library search. Expert programs should be used after you have studied the spectrum in question and understand as much about it as possible. The more informed you are about a spectrum, the more accurately you will interpret the output of an expert interpretation program.

VI. Infrared Interpretation and the Internet

It is always dangerous to report on a phenomenon that is changing as fast as the internet. Things on the internet that were once thought impossible are now possible and will be commonplace tomorrow. Thus, this section runs the risk of becoming dated rather quickly. However, the development of the internet promises to have a huge impact on many worlds, including that of infrared spectroscopy, which is why this topic is worth discussing.

A search of the World Wide Web using any of the standard search engines with phrases such as "infrared spectral interpretation" or "infrared spectral libraries" will many matches. Many of these matches will not be of interest to the typical chemical analyst. Therefore, we will discuss here the types of web sites that most commonly have useful information.

Many colleges and universities have begun to put course contents, exercises, laboratories, and exams on the internet. Several universities have small collections of spectra or spectral unknowns that can be used to challenge beginning interpreters. These sites can be useful sources of information and exercises for beginning infrared interpreters.

For those of you with access to the Usenet newsgroups, there is a bulletin board called "sci.techniques.spectroscopy" where all sorts of questions about spectroscopy are asked and answered. This newsgroup is particularly useful for finding out obscure or difficult to find information. Thousands of spectroscopists read this newsgroup every day, and chances are one of them has the answer to your question.

A unique feature of the internet is the existence of infrared spectral libraries that can be downloaded and used free of charge. Once these spectra are placed in a library, they can assist in the process of identifying unknown spectra. The website of Galactic Industries (www.galactic.com) has links to several free collections of infrared spectra. Some of the spectra in this book were obtained in this fashion.

There are some problems, however, that may be encountered in taking spectra off the internet. First, read the fine print on the web site about what uses are allowed for the data. For example, you may be forbidden to use the data for "commercial purposes." Second, be skeptical about the quality of the data. Look the spectra over carefully, make sure they are noise free, well resolved, and that they are free of artifacts. You may want to go as far as e-mailing the webmaster to ascertain the source of the spectra, and to gain more information such as the source of the sample, the instrument used, sampling techniques, etc. Last, make sure the sample is properly identified. If a spectrum claims to be that of a polymer, but looks more like the spectrum of an inorganic material, be suspicious. Remember, by using someone else's data you are trusting their judgement and competence in taking a good spectrum and properly identifying it. Make sure the spectrum looks correct to your eyes before downloading it and using it.

As time goes on and the cost of transporting information across the internet falls to near zero, scientists will increasingly begin to post their own collection of spectra on the internet. The day may come when entire web sites full of infrared spectral libraries will exist, and all one has to do to search these databases is upload your spectrum to the site, perform the search, then download the results. Commercial infrared libraries may become obsolete.

A version of this "spectral library searching on the internet" already exists. There are companies that will, for a fee, take a spectrum you e-mail to them, perform a spectral library search for you, then e-mail the results back to you. Some of these firms claim a 24-hour turn around time for these services. If you have only occasional need for spectral library searching, this may be a viable option.

Last, the web sites of the major FTIR instrument manufacturers and scientific software companies have information about books, training courses, and software packages that can act as interpretation aids. The author invites you to browse his company web site at www.spectros-associates.com for the latest news on infrared spectral interpretation course offerings.

Bibliography

B.C. Smith, *Fundamentals of Fourier Transform Infrared Spectroscopy*, CRC Press, Boca Raton, 1996.

Appendix 1

Answers to Problem Spectra

On the following pages, the correct interpretation of the problem spectra found at the end of Chapters 2 through 5 are given. The name, chemical structure, and chemical formula of the molecule that gave rise to the spectrum will be given. A table containing the assignment of the important peaks in each spectrum will also be given. A discussion of the proper strategy to follow to interpret each spectrum is included. The limitations of each spectrum and how it may have mislead or confused the reader will also be discussed.

Please realize that the complete chemical structure of a molecule cannot always be determined from the infrared spectrum alone. However, much can be learned from a spectrum. The point of these exercises is to teach readers the process of how to interpret a spectrum, the types of information that can be gleaned from a spectrum, and the limitations of infrared spectroscopy as a chemical analysis technique. Once these ideas are understood, you are well on your way to properly using infrared spectra.

The answers are grouped in the order they appear in the book. So, we start with the first problem in Chapter 2, and end with the last problem in Chapter 5. Not all chapters had problem spectra included in them. I hope you find these exercises useful.

Problem 2.1

Polyethylene $(CH_2)_n$

Peak	Assignment
2936	CH_2 Asymmetric stretch
2855	CH_2 Symmetric Stretch
1464	CH_2 Scissors
719	CH_2 Rock

This is perhaps one of the simplest spectra you will ever see. There are only four peaks of note, all of which can be assigned to the methylene group. The C-H stretches at 2936 and 2855 cm^{-1} firmly establish the presence of CH_2 in this molecule. The fact that there are only two C-H means there is only one kind of saturated hydrocarbon functional group present. Once you know there are methylenes present, you should attempt to find how many methylenes there are in a row. Do this by looking for the CH_2 rock near 720 cm^{-1}. This band is present, indicating that there are at least four CH2s in a row. The band at 1460 cm^{-1} may be due to a methyl or methylene group, and so is not very useful.

The <u>lack</u> of any methyl bands is important. Methylene chains usually end in a CH_3, yet in this spectrum there is a chain of methylenes without a methyl group. This leaves two possibilities, either the methylene chain is "infinitely long" or the molecule is a ring of CH2s. The fact that this sample is a solid argues for the former. Most saturated, cyclic hydrocarbons are liquids at room temperature. Additionally, cyclic alkanes like cyclohexanes have more complicated spectra than this. Thus, the molecule is polyethylene.

Problem 2.2

CH$_3$-CH$_2$-CH$_2$-CH$_2$-CH$_2$-CH$_2$-CH$_2$-CH$_3$

Octane (C$_8$H$_{18}$)

Peak	Assignment
2959	CH$_3$ Asymmetric stretch
2926	CH$_2$ Asymmetric stretch
2876	CH$_3$ Symmetric stretch
2858	CH$_2$ Symmetric stretch
1379	CH$_3$ Umbrella mode
722	CH$_2$ Rock

This spectrum is very similar to that of n-hexane (Figure 2.4) and is a classic example of a saturated hydrocarbon. The series of four peaks between 2800 and 3000 cm^{-1} are easily assigned as the stretching vibrations of methyl and methylene groups. The umbrella mode confirms the presence of a methyl group in the sample. The umbrella mode is not split, indicating there are no branch points in this molecule, and that this is a straight chain alkane.

Once the presence of methylene is suspected, you should look for the CH$_2$ rock near 720 cm^{-1}. It is present in this spectrum, indicating that there are at least 4 CH2s in a row. As mentioned in Chapter 2, the relative intensity of the CH$_3$ and CH$_2$ asymmetric stretches is related to the length of a hydrocarbon chain. Notice that in the spectrum of hexane, these two bands are of about equal intensity. In Problem 2.2, the CH$_2$ stretch is more intense than the CH$_3$ stretch, indicating that this hydrocarbon chain is longer than the one in hexane. However, comparison to the spectrum of vaseline in Figure 2.8 shows that this sample's chain is shorter than that of vaseline. Thus, we can say there are somewhere between 4 and 18 methylenes in a row. However, the spectrum looks more like hexane than vaseline, so the number of methylenes is probably closer to four than to eighteen. Unfortunately, there is no sure way to determine the exact number of methylenes in this molecule. Therefore, the best conclusion to be drawn from this spectrum is that it is a straight chain saturated hydrocarbon somewhat longer than hexane. The exact answer is octane.

Problem 2.3

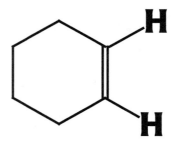

Cyclohexene (C_6H_{10})

Peak	Assignment
3022	Alkene CH stretch
2926	CH_2 Asymmetric stretch
2837	CH_2 Symmetric stretch
1652	Alkene C=C stretch
718	Alkene out-of-plane C-H bend
640	Alkene out-of-plane C-H bend

The C-H stretches above and below 3000 cm^{-1} indicate that this molecule has saturated and unsaturated portions. Once unsaturation is suspected, the type of unsaturation should be determined by looking near 1600 cm^{-1}. The C=C stretching band at 1652 cm^{-1} confirms the existence of the C=C, indicating that the double bond is vinyl, vinylidine, or cis. The next thing to determine is the substitution pattern around the C=C. Vinyl and vinylidine groups contain a =CH_2 group, which has an asymmetric stretch between 3090 and 3075 cm^{-1}. The unsaturated C-H stretches in this spectrum do not fall in this range, indicating the molecule probably has cis geometry. The high intensity, low wavenumber out-of-plane C-H bending vibrations must be consulted to confirm the structure. There are no bands at the correct wavenumber to be vinyl or vinylidine groups. However, there are bands at 718 and 640 cm^{-1} that fall into the range of the out-of-plane C-H bend for a cis double bond. This confirms that the double bond has a cis geometry, and that there are two substituents on the double bond. The next task is to determine the nature of these substituents. The saturated C-H stretches indicate the presence of methylene groups, but no strong indication of methyl. Additionally, there is no umbrella mode, confirming the absence of methyl groups. Like Problem 2.1, we are again faced with the problem of how to end a methylene chain. Essentially, we must have either an "infinite" chain or a cyclic compound. The fact that this sample is a liquid argues that it is a cyclic compound consisting of methylenes and a cis double bond. Organic chemistry tells us that carbon atoms most commonly form 5 or 6 membered rings. This narrows it down to two compounds, cyclopentene or cyclohexene. The correct answer is cyclohexene.

Normally, once the presence of methylene groups is suspected, you should go in search of the CH_2 rock at 720 cm^{-1} to try to determine how many methylenes there are in a row. In the spectrum of cyclohexene, there is a band conveniently located at 718 cm^{-1}, which you might suspect as being a CH_2 rock. However, 4 methylenes in a ring do not exhibit the 720 cm^{-1} band, as can be seen from an examination of the spectra of cyclohexane and cyclohexanol. In this instance, the band at 718 cm^{-1} is one of the out-of-plane C-H bending bands of the cis double bond. Fortuitously, cyclohexene does have 4 CH_2s in a row and in this instance a misassignment of the 718 cm^{-1} band is not fatal. However, it could be a problem with other cyclic alkenes.

Problem 2.4

$$\text{CH}_3\text{CH}_2\text{CH}_2\text{CH}_2-\underset{H}{\overset{H}{C}}=\underset{H}{\overset{H}{C}}$$

1-Hexene (C_6H_{12})

Peaks	Assignments
3081	Alkene =CH$_2$ stretch
2962	CH$_3$ Asymmetric stretch
2930	CH$_2$ Asymmetric stretch
2875	CH$_3$ Symmetric stretch
2863	CH$_2$ Symmetric stretch
1643	Alkene C=C stretch
1380	CH$_3$ Umbrella mode
993	Alkene out-of-plane C-H bend
910	Alkene out-of-plane C-H bend

    This sample has both saturated and unsaturated portions as indicated by C-H stretches above and below 3000 cm^{-1}. The next step is to determine the type of unsaturation. An examination of the 1600 cm^{-1} region shows a band at 1643 cm^{-1}, which is the C=C stretch of a vinyl, vinylidine, or cis double bond. The unsaturated C-H stretch at 3081 cm^{-1} is in position to be the asymmetric stretch of the =CH$_2$ group. This would narrow the choices to a vinyl or vinylidine group. To determine the substitution pattern around the double bond, a search for the out-of-plane C-H bending bands must be made. There are strong bands at 993 and 910 cm^{-1}, which are due to a vinyl group.

    The next job is to determine the identity of any substituents attached to the double bond. By definition, vinyl groups have just one substituent attached to the C=C. The saturated C-H stretches indicate the presence of methyl and methylenes in the sample. The umbrella mode at 1382 cm^{-1} confirms the presence of a methyl group, and since it is not split means this is a straight chain alkane. A search for the CH$_2$ rocking band at 720 cm^{-1} turns up nothing, meaning that there are less than 4 methylenes in a row in this molecule. Thus, the substituent must be a saturated hydrocarbon chain with 1, 2, or 3 methylenes in it. The molecule is 1-hexene.

Problem 2.5

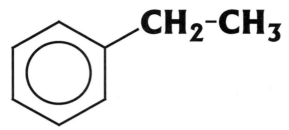

Ethylbenzene (C_8H_{10})

Peak	Assignment
3084	Aromatic C-H stretch
3063	Aromatic C-H stretch
3027	Aromatic C-H stretch
2968	CH_3 Asymmetric stretch
2931	CH_2 Asymmetric stretch
2873	CH_3 Symmetric stretch
1604	Ring Mode
1495	Ring Mode
1375	CH_3 Umbrella mode
745	Aromatic out-of-plane C-H bend
697	Ring bend

The C-H stretches above and below 3000 cm^{-1} indicate this molecule has saturated and unsaturated portions. To determine the type of unsaturation, look around 1600 cm^{-1}. There are no bands between 1630 or 1680 cm^{-1} that would indicate an alkene. However, strong, sharp bands at 1604 and 1495 cm-1 are the ring modes of an aromatic ring. The next step is to determine the substitution pattern on the benzene ring. The out-of-plane C-H bend at 745 cm^{-1} and the ring bend at 697 cm^{-1} firmly indicate a monosubstituted benzene ring.

The last task, then, is to determine the nature of the substituent on the benzene ring. The saturated C-H stretches show there is a methyl group present because of the asymmetric and symmetric stretches at 2968 and 2873 cm^{-1}, and the umbrella mode at 1375 cm^{-1}. The well-defined methylene asymmetric stretch at 2931 cm^{-1} indicates there is a CH_2 in the sample. The CH_2 symmetric stretch is not present, and may be obscured by other nearby bands. This hydrocarbon chain is not branched because the umbrella mode is not split.

To determine the number of methylenes in a row, one would normally look for the CH_2 rocking vibration at 720 cm^{-1}. However, this region is dominated by intense benzene ring bands, and it is hard to tell whether or not the rocking band is present. In this case, the best way to determine hydrocarbon chain length is to compare the intensity of the asymmetric stretches of the methyl and methylene groups. The methyl band is much more intense than the methylene band, indicating the substituent is a short hydrocarbon chain. However, it is difficult to determine the precise number of methylene groups. The molecule is ethylbenzene, which contains only one CH_2 group.

Problem 2.6

Isopropylbenzene (C$_9$H$_{12}$)

Peak	Assignment
3085	Aromatic C-H stretch
3064	Aromatic C-H stretch
3029	Aromatic C-H stretch
2962	CH$_3$ Asymmetric stretch
2872	CH$_3$ Asymmetric stretch
1604	Aromatic ring mode
1494	Aromatic ring Mode
1384, 1364	Split CH$_3$ umbrella mode, 1:1 intensity
760	Aromatic out-of-plane C-H bend
699	Aromatic ring bend

C-H stretches above and below 3000 cm^{-1} indicate saturated and unsaturated portions exist in this molecule. An examination of the 1600 cm^{-1} region reveals no features between 1680 and 1630 cm^{-1}, meaning the molecule does not contain a double bond. Sharp features at 1604 and 1494 cm^{-1} are ring modes of an aromatic ring. To determine the ring substitution pattern, the out-of-plane C-H bending vibrations between 860 and 700 cm^{-1} must be found. There is a C-H bend at 760 cm^{-1} and a ring bend at 699 cm^{-1}. These two bands mean the molecule could be mono- or meta- substituted.

The saturated portions of the molecule must make up the benzene ring substituent(s). The saturated C-H stretches show CH$_3$, but no CH$_2$ present. The methyl group umbrella mode is split, with halves of about equal intensity at 1384 and 1364 cm^{-1}. This indicates there is a branch point in the molecule, and that it is either isopropyl or gem-dimethyl.

The simplicity of the spectrum and the fact that all the major bands have already been assigned, argues for a single ring substituent. If there are two substituents, they must be identical. The molecule is isopropylbenzene.

Problem 2.7

CH$_3$–(CH$_2$)$_{16}$–CH$_3$

Octadecane (C$_{18}$H$_{38}$)

Peak	Assignment
2956	CH$_3$ Asymmetric stretch
2922	CH$_2$ Asymmetric stretch
2858	CH$_2$ Symmetric stretch
1381	CH$_3$ Umbrella mode
721	CH$_2$ Rocking

With no C-H stretches above 3000 cm^{-1}, this is a totally saturated molecule. The molecule must consist strictly of methyl, methylene, and methine groups. The C-H stretches show strong CH$_2$ asymmetric and symmetric stretching at 2922 and 2858 cm^{-1}. There is a small shoulder corresponding to a methyl asymmetric stretch at 2956 cm^{-1}, but the best indicator of a methyl group is the umbrella mode at 1381 cm^{-1}. The umbrella mode is not split, meaning this is probably a straight chain alkane. To determine the number of methylenes in a row, search for the CH$_2$ rocking vibration. It is found at 721 cm^{-1}, indicating that there are at least 4 CH2s in a row. The great intensity of the CH$_2$ asymmetric stretch compared to the CH$_3$ asymmetric stretch indicates this is a very long hydrocarbon chain. In fact, this spectrum is very similar to that of vaseline. Unfortunately, we cannot determine the exact hydrocarbon chain length from the spectrum. The molecule is octadecane, a straight chain alkane with 18 carbon atoms.

Problem 2.8

Tert-butylbenzene ($C_{10}H_{14}$)

Peak	Assignment
3089	Aromatic C-H stretch
3062	Aromatic C-H stretch
3029	Aromatic C-H stretch
2965	CH_3 Asymmetric stretch
2870	CH_3 Asymmetric stretch
1602	Aromatic ring mode
1497	Aromatic ring mode
1394, 1366	Split CH_3 umbrella mode, 1:2 intensity ratio
763	Aromatic out-of-plane C-H bend
698	Aromatic ring bend

As in many of the other problem spectra, the C-H stretches above and below 3000 cm^{-1} indicate there are saturated and unsaturated portions in this molecule. There are no C=C stretching bands between 1680 and 1630 cm^{-1}, but there are sharp benzene ring modes at 1602 and 1497 cm^{-1}. Therefore, the unsaturation type is an aromatic ring. Examination of the low wavenumber part of the spectrum shows intense bands at 763 and 698 cm^{-1}, which are assigned as the out-of-plane C-H bend and ring bend, respectively. These two bands together indicate the molecule is either mono- or meta-substituted.

Strong unsaturated C-H stretches at 2965 and 2870 cm^{-1} indicate a methyl group is present, but there is no CH_2 stretching band present. Close examination of the methyl group umbrella mode will show it is split in two, with halves of unequal intensity at 1394 and 1366 cm^{-1}. This split umbrella mode is indicative of a branch point, and the 1:2 intensity ratio of these bands makes it either a tert-butyl or an isobutyl branch point. The simplicity of the spectrum, and the fact that all the major peaks have already been assigned, argues that it is mono- rather or symmetrically meta- substituted. The molecule is tert-butylbenzene.

Problem 3.1

CH₃CH₂CH₂-OH

n-Propanol (C$_3$H$_8$O)

Peak	Assignment
3342	O-H Stretch
2965	CH$_3$ Asymmetric stretch
2938	CH$_2$ Asymmetric Stretch
2879	CH$_3$ Symmetric stretch
1385	CH$_3$ Umbrella mode
1058	Primary alcohol C-O stretch
969	C-C-C Stretch
889	C-CH$_2$-O Symmetric stretch
660	O-H Out-of-plane bend

The large O-H stretch at 3342 cm^{-1} indicates that this molecule is an alcohol. It is good practice after spotting the O-H stretch to then track down the O-H bending vibrations. The O-H in-plane bend should appear at 1350±50 cm^{-1}. There is a broad envelope in this vicinity with sharper C-H bending bands superimposed on it that is probably the O-H in-plane bend. The O-H out-of-plane bend is visible at 660 cm^{-1}, and is easy to spot because of its width.

The C-H stretches indicate this is a totally saturated molecule. The methyl group asymmetric and symmetric C-H stretches are at 2965 and 2879 cm^{-1}. The methylene asymmetric stretch appears clearly at 2938 cm^{-1}, but is less intense than the methyl asymmetric stretch. This indicates that the saturated hydrocarbon chain in this molecule is short. To determine the number of CH2s in a row, look for the CH$_2$ rocking band at 720 cm^{-1}. There is no band in this region, so the hydrocarbon chain must contain 3 or fewer methylenes in a row. If the 720 cm^{-1} band were present, it would appear as a sharp band superimposed on the O-H bend at 660 cm^{-1}.

There is a single umbrella mode at 1385 cm^{-1} indicating there are probably no branch points in this molecule. The last thing we need to determine is the type of alcohol: primary, secondary, or tertiary. Remembering that the C-O stretch is usually the most intense band between 1300 and 1000 cm^{-1}, we would assign the band at 1058 cm^{-1} as being the alcohol C-O stretch. This is then a primary alcohol. Remember that a primary alcohol contains a -CH$_2$-OH group by definition.

Thus, we have a primary alcohol with no branch points and less than 4 methylenes in a row. This narrows it down to n-pentanol, n-butanol, n-propanol, or ethanol (whose spectrum is seen in Figure 3.5). This points out a limitation of infrared spectra, the difficulty of determining the exact number of methylenes in a hydrocarbon chain. The best way of distinguishing between the alcohols just listed would be to consult an atlas of infrared spectra or perform a library search. The compound is n-propanol.

Problem 3.2

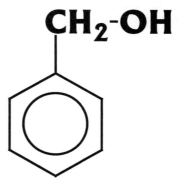

Benzyl Alcohol (C$_7$H$_8$O)

Peak	Assignment
3334	O-H Stretch
3029	Aromatic C-H stretch
2930	CH$_2$ Asymmetric stretch
2872	CH$_2$ Symmetric stretch
1495	Aromatic ring mode
1453	Aromatic ring mode
1019	Primary alcohol CH$_2$-O stretch
736	Aromatic out-of-plane C-H bend
697	Aromatic ring bend

The strong, broad band at 3334 cm^{-1} is an O-H stretch of an alcohol. The O-H bending bands of the alcohol should next be found, but they are difficult to see in this spectrum. The in-plane O-H bend should be at 1350±50 cm^{-1}. In this region, there is a broad envelope with sharper bands superimposed on it. This envelope is the O-H bend. A similar situation occurs for the out-of-plane O-H bend. Normally found at 650±50 cm^{-1}, this band can be seen underneath a number of sharp bands superimposed upon it.

The C-H stretch region shows that this is a mixed molecule, with unsaturated and saturated portions. Examination of the 1680-1630 cm^{-1} region shows no C=C bands, whereas sharp, intense aromatic ring modes at 1495 and 1453 cm^{-1} show that the source of unsaturation is a benzene ring. The next question to answer is the substitution pattern on the benzene ring. A search for the out-of-plane C-H bending vibration turns up a strong band at 736 cm^{-1} accompanied by a ring-bending band at 697 cm^{-1}. These bands firmly indicate the benzene ring is monosubstituted.

Next, we need to determine the type of alcohol. Since there is a benzene ring present, it is possible the O-H is directly attached to the ring, giving a phenol. However, a search for the C-O stretch indicates the strongest band between 1300 and 1000 cm^{-1} is at 1019 cm^{-1}, making this a primary alcohol. Remember that a primary alcohol must have a methylene group in it (-CH$_2$OH).

The saturated C-H stretches are a little misleading. The band at 2930 cm^{-1} could be assigned as a methylene asymmetric stretch, while the band at 2872 cm^{-1} is in position to be a CH$_3$ symmetric stretch. However, note the pattern of bands here. There are only two saturated C-H stretches, indicating there is just methyl or just methylene in this molecule. Secondly, the band assigned as a symmetric stretch is more intense than the band assigned as an asymmetric stretch. This defies the normal pattern in a hydrocarbon and indicates there is a saturated carbon attached to an oxygen atom. We already know there is a -CH$_2$-O- because we have a primary alcohol. Examination of Table 3.3 shows the bands at 2930 and 2872 cm^{-1} fall into the ranges for a CH_2 bonded to an oxygen atom. Therefore, these bands are consistent with our assignment of a primary alcohol.

Since the benzene ring has only one substituent, the primary alcohol must be attached to the ring. The only question left is how many methylenes are there in the chain that is part of the primary alcohol. Normally, a search for the CH_2 rocking band at 720 cm^{-1} would be made. However, the strong, sharp benzene ring bands in this region make it impossible to determine whether or not there is a CH_2 rocking band present. Again, being able to determine the exact number of methylenes in a hydrocarbon chain is not possible. The molecule is benzyl alcohol, which contains just one CH_2 group.

What about the band at 1369 cm^{-1}? This band looks like a methyl group umbrella mode. However, as noted above the number and position of the saturated C-H stretches, combined with the knowledge that we have a primary alcohol, argues against there being a methyl group in this sample. The band at 1369 cm^{-1} is best assigned as a ring mode.

Problem 3.3

t-Butanol ($C_4H_{10}O$)

Peak	Assignment
3371	O-H Stretch
2974	CH_3 Asymmetric Stretch
1379, 1366	Split CH_3 umbrella mode, 1:2 intensity
1202	Tertiary C-O stretch
914	C-C-C Stretch
648	Out-of-plane O-H bend

The strong, broad O-H stretch at 3371 cm^{-1} means this molecule is an alcohol. The O-H bends for the alcohol would be expected to appear at 1350±50 and 650±50 cm^{-1}. The sharp bands near 1350 cm^{-1} are superimposed on an envelope that is probably the O-H in-plane bend. The out-of-plane O-H bend is clearly seen at 648 cm^{-1}. The C-H stretches show that this molecule is totally saturated. To ascertain the type of alcohol, we need to look for the most intense band between 1300 and 1000 cm^{-1}. It falls at 1202 cm^{-1}. This is in position to be either a tertiary alcohol or a phenol. However, since this molecule is totally saturated, there cannot be an aromatic ring present, so it must be a tertiary alcohol.

The final step of the analysis is to figure out what three substituents are attached to the hydroxyl carbon. The C-H stretches show only one strong band at 2974 cm^{-1}, which is the methyl group asymmetric stretch. There is no indication of methylenes in this spectrum. Since a tertiary alcohol contains a branch point, examination of the umbrella mode may tell us what type of branch point. The umbrella mode is split in two, with bands of uneven intensity at 1379 and 1366 cm^{-1}. This indicates the branch point is either t-butyl or isobutyl. However, isobutanol is a primary alcohol (see the structure of the isobutyl group in Chapter 2), thus, the molecule must be t-butanol.

Problem 3.4

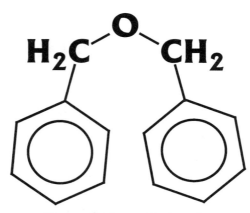

Benzyl Ether ($C_{14}H_{14}O$)

Peak	Assignment
3088, 3064, 3031	Aromatic C-H stretches
2924	CH_2 Asymmetric stretch
2858	CH_2 Symmetric stretch
1604, 1586, 1497, 1362	Aromatic ring modes
1095	Saturated ether C-O stretch
737	Aromatic out-of-plane C-H bend
697	Aromatic ring bend

The C-H stretches above and below 3000 cm^{-1} means this is a mixed molecule with saturated and unsaturated portions. An examination of the middle of the spectrum shows no C=C stretches between 1680 and 1630 cm^{-1}, but there are aromatic ring modes at 1604, 1586, 1497, and 1362 cm^{-1}. To determine the benzene ring substitution pattern, we must look at the lower wavenumber region of the spectrum. There is a clear out-of-plane C-H bend at 737 cm^{-1} and a ring-bend at 697 cm^{-1}. These bands are a clear indication that this is a monosubstituted benzene ring.

There are only two saturated C-H stretch bands, indicating that the saturated portions of the molecule are either all methyl or all methylene. The bands at 2924 and 2858 cm^{-1} are the asymmetric and symmetric stretches of a methylene group. These bands fall in the expected range for methylene C-H stretches. However, the symmetric stretch is _more_ intense than the asymmetric stretch, which should immediately make you think the CH_2 may be attached to an oxygen atom. Ideally, we would look for the CH_2 rocking band at 720 cm^{-1} to ascertain the number of methylenes in a row. Unfortunately, the region is dominated by benzene ring bands, and it is impossible to tell whether the CH_2 rocking band is present or not. There is no methyl umbrella mode, and no indication in the C-H stretch region of a methyl group.

The position and intensity of the band at 1095 cm^{-1} makes it a good candidate for a C-O stretch. Since there is no O-H group present, this band must be due to an ether. Its position is in the right place for a saturated ether. The fact that there is only one strong band between 1000 and 1300 cm^{-1} indicates that there is no branching on the ether carbons and that there are CH_2 groups attached to both sides of the oxygen. So, the pieces of the puzzle are a monosubstituted benzene ring, a methylene, and a saturated ether. The CH2s must be attached to the oxygen for the ether to be saturated, and they must be attached to the benzene ring for there to be just one substituent. Thus, the only detail we cannot determine is the _number_ of methylenes between the benzene ring and the oxygen. It turns out there is only 1 CH_2 on each side of the oxygen and the molecule is benzyl ether.

Problem 3.5

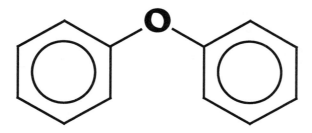

Phenyl Ether (C$_{12}$H$_{10}$O)

Peak	Assignment
3068	Aromatic C-H stretch
3041	Aromatic C-H stretch
1585	Aromatic ring mode
1488	Aromatic ring mode
1238	Aromatic ether C-O stretch
750	Out-of-plane C-H bend
692	Aromatic ring bend

The C-H stretches above 3000 cm^{-1}, combined with the lack of C-H stretches below 3000 cm^{-1}, indicates this molecule is totally unsaturated. To determine the type of unsaturation, we look at the 1600 cm^{-1} region. There are no bands between 1680 and 1630 cm^{-1}, but there are strong, sharp bands at 1585 and 1488 cm^{-1}. These are the ring modes of a benzene ring. To determine the substitution pattern on the benzene ring, look for the out-of-plane C-H bend at low wavenumber. It appears at 750 cm^{-1} accompanied by a ring bend at 690 cm^{-1}. These bands together indicate that the benzene ring is monosubstituted.

The intense band at 1238 cm^{-1} is in the right place to be a C-O stretch and, since there are no O-H bands present; this must be an ether. The band is in the right place to be an aryl ether and, since the molecule is totally unsaturated, this makes sense. The fact that there is only one strong C-O stretch means the ether is probably symmetrical. Additionally, since our benzene ring is mono-substituted, the ring must be directly attached to the ether oxygen. The molecule is phenyl ether.

Problem 3.6

$$CH_3\diagup O \diagdown CH_2CH_2CH_2CH_3$$

Butyl Methyl Ether ($C_5H_{12}O$)

Peak	Assignment
2962	CH_3 Asymmetric stretch
2934	CH_2 Asymmetric stretch
2871	CH_3 Symmetric stretch
2829	CH_3-O Symmetric stretch
1386	CH_3 Umbrella mode
1124	Ether C-O stretch
830	Symmetric C-C-O stretch

With C-H stretches below 3000 cm^{-1}, this is a totally saturated molecule. The methyl asymmetric and symmetric stretches show up at 2962 and 2871 cm^{-1}, and the methylene asymmetric stretch is seen at 2934 cm^{-1}. To ascertain the number of methylenes in a row, look for the CH_2 rock at 720 cm^{-1}. There is no band in that vicinity, so the hydrocarbon chains in the molecule must be short. The CH_3 asymmetric stretch is more intense than the CH_2 asymmetric stretch, also indicating short chains. The methyl group umbrella mode is clearly seen at 1386 cm^{-1}. Since it is not split, we have straight hydrocarbon chains.

The intense band at 1124 cm^{-1} is the C-O stretch of a saturated ether. The only challenge left is to figure out the exact length of the hydrocarbon chains hanging off the oxygen. The low wavenumber medium intensity C-H stretch at 2829 cm^{-1} is the symmetric stretch of a methoxy group. This immediately tells us that one of the ether substituents is a methyl group. The lack of a split umbrella mode and the one intense band between 1000 and 1300 cm^{-1} indicates there is no branching on the second ether carbon. The ether linkage must be -CH_2-O-CH_3. This narrows it down to methyl ethyl, methyl propyl, or methyl butyl ether. Unfortunately, we cannot determine the exact number of methylenes. The molecule is n-butyl methyl ether, an asymmetric saturated ether.

Problem 3.7

Hydroquinone or p-Hydroxy Phenol ($C_6H_6O_2$)

Peak	Assignment
3260	O-H Stretch
3018	Aromatic C-H stretch
1464	Aromatic ring mode
1346	In-plane O-H bend
1205	Phenol C-O stretch
759, 828	Aromatic out-of-plane C-H bends
610	O-H Out-of-plane bend

The strong, broad O-H stretch at 3260 cm^{-1} clearly indicates this is an alcohol. The O-H in-plane and out-of-plane bending vibrations are at 1346 and 610 cm^{-1}, respectively. The sharp band at 3018 cm^{-1} is superimposed on the much broader O-H stretch is an unsaturated C-H stretch. There are no saturated C-H stretches below 3000 cm^{-1}, so this molecule is totally unsaturated.

The type of unsaturation can be determined by looking near 1600 cm^{-1}. There are no C=C bands between 1680 and 1630, but there are aromatic ring modes at 1505 and 1464 cm^{-1}. The band at 1346 cm^{-1} is <u>not</u> a methyl umbrella mode because this molecule is totally unsaturated. To determine the benzene ring substitution pattern, we must look for the out-of-plane C-H bends at low wavenumber. There are two bands of equal intensity at 828 and 759 cm^{-1}, both of which are C-H bends. There is no band at 690 cm^{-1}, meaning the substitution pattern is ortho or para. At a minimum, this tells us that the benzene ring is disubstituted. Unfortunately, the two C-H bends fall into positions where they could be ortho or para, making it impossible to determine the exact substitution pattern.

To determine the type of alcohol present, we need to look for the most intense band between 1300 and 1000 cm^{-1}. The asymmetric C-C-O stretch occurs in this spectrum at 1205 cm^{-1}. This is in a range to be either a phenol or a saturated tertiary alcohol. However, since the C-H stretches indicate this molecule is totally unsaturated, we must have a phenol.

Having identified one benzene ring substituent, we need to determine the other. However, there are no other major bands in the spectrum and the simplicity of the spectrum argues for the two substituents being the same. That happens to be the case, and the molecule is hydroquinone (common name) or p-hydroxy phenol.

Problem 3.8

HO—⟨○⟩—O—CH₃

p-Methoxy Phenol or p-Hydroxy Anisole ($C_7H_8O_2$)

Peak	Assignment
3384	O-H Stretch
3078, 3027	Aromatic C-H stretches
2950	CH$_3$ Asymmetric stretch
2831	CH$_3$-O Symmetric stretch
1630, 1604	Aromatic ring modes
1508	Aromatic ring mode
1354	O-H In-plane bend
1222	Aromatic ether and phenol C-O stretch
1025	Saturated ether C-O stretch
821	Aromatic out-of-plane C-H bend

The strong, broad O-H stretch at 3384 cm^{-1} clearly indicates this is an alcohol. The O-H in-plane bending band appears at 1354 cm^{-1}, while the out-of-plane O-H bend is weak. The C-H stretches above 3000 cm^{-1} indicate unsaturation, and the ring modes such as the ones at 1604 and 1508 cm^{-1} disclose that the unsaturation is a benzene ring. To ascertain the substitution pattern on the ring, the out-of-plane C-H bends of the ring must be found. There is a strong band at 821 cm^{-1} that fits the bill. Note that there is <u>no</u> band at 690 cm^{-1}. This means the benzene ring is para substituted.

There are only two saturated C-H stretches, indicating there is just methyl or just methylene in the molecule. The band at 2960 cm^{-1} is easily assigned as the CH$_3$ asymmetric stretch. The band at 2831 cm^{-1} is much too low to be a normal CH$_3$ symmetric stretch. However, it is perfectly placed to be the symmetric stretch of an O-CH$_3$ group (see Table 3.3). This immediately tells us that there is a methoxy group in the molecule in addition to an alcohol. The umbrella mode for a methoxy group does not show up around 1375 cm^{-1}, which is why the band at 1354 cm^{-1} is best assigned as the O-H bend.

Armed with the knowledge that we have an alcohol and ether linkage, it is now time to make sense of the C-O stretching region between 1300 and 1000 cm^{-1}. There are two intense bands in this region, 1222 and 1025 cm^{-1}. The band at 1025 cm^{-1} could be due to a primary alcohol, -CH$_2$-OH. However, the C-H stretches have already shown there are no methylenes in this molecule, so there cannot be a primary alcohol. We know we have an O-CH$_3$ linkage, and its C-O stretch is the band at 1025 cm^{-1}. The C-O stretch at 1222 cm^{-1} can be either a phenol or aromatic ether linkage. Since there are no saturated groups present but an O-CH$_3$, the O-H must be attached to the benzene ring giving us a phenol. Again, since the CH$_3$ is the only saturated moiety present, the benzene ring must also be directly attached to the ether oxygen, giving an aryl/alkyl ether. We know the substituents are para substituted, so the molecule must be p-methoxy phenol.

Problem 4.1

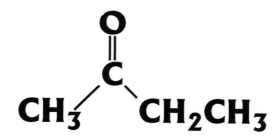

Methyl Ethyl Ketone or "MEK" (C$_4$H$_8$O)

Peak	Assignment
2979, 2882	Methyl C-H stretches
2939	Methylene C-H stretch
1717	C=O Stretch
1366	CH$_3$ Umbrella mode
1630, 1604	Aromatic ring modes
1172	Ketone C-C-C stretch

The C-H stretches below 3000 cm^{-1} indicate this molecule is totally saturated. There are 4 well-defined C-H stretching bands, indicating the molecule probably contains methyl and methylene groups. The positions of the C-H stretches fall a few wavenumbers outside their accepted ranges, but are close enough for a tentative assignment. The strong band at 1717 cm^{-1} is a carbonyl C=O stretching vibration. This band does not tell us to which functional group the carbonyl belongs.

The molecule is not an aldehyde because there is not a low wavenumber C-H stretch. The molecule is not an acid because there is not a strong O-H stretching vibration, and it is not an ester because it does not have the rule of three bands. This leaves us with a saturated ketone. Such a molecule should have a C-C-C stretch between 1230 and 1100 cm^{-1}, and the band at 1172 cm^{-1} establishes this molecule as a saturated ketone.

The last task is to determine what saturated hydrocarbon chains are attached to the carbonyl carbon. The molecule definitely has methyl groups because of the C-H stretches and the umbrella mode at 1366 cm^{-1}. The four C-H stretches indicate that there are probably also some methylenes in the molecule. Examination of the 720 cm^{-1} region shows no CH$_2$ rocking vibration present, so the methylenes present are in chains of less than four in a row. The lack of splitting in the umbrella mode indicates both hydrocarbon chains are unbranched. Thus, both substituents are straight chain hydrocarbons with less than 4 methylenes in each chain. This narrows us to a small family of saturated ketones, but the spectrum does not allow us to nail down the exact number of methylenes, so we cannot completely identify the molecule. The molecule is methyl ethyl ketone.

Problem 4.2

p-Diacetyl Benzene ($C_{10}H_{10}O_2$)

Peak	Assignment
3076	Aromatic C-H stretch
2990	Methyl C-H stretch
1679	C=O Stretch
1449, 1423	Aromatic ring modes
1375	CH_3 Umbrella mode
1250	Ketone C-C-C stretch
829	Aromatic out-of-plane C-H bend

The C-H stretches at 3079 and 2990 cm^{-1} indicate this molecule has saturated and unsaturated portions. The unsaturation could be a C=C bond or a benzene ring. An alkene has a C=C stretching band between 1630 and 1680 cm^{-1}, and the band at 1679 cm^{-1} could be assigned as such. However, this band is much too intense to be a C=C stretch. Additionally, there are what appear to be aromatic ring modes at 1449 and 1423 cm^{-1}, so the source of unsaturation is probably an aromatic ring.

There appears to be only one or two saturated C-H stretches. This means there is either only methyl or methylene in the molecule. The high wavenumber of the C-H stretch argues for the saturated moiety to be a methyl group. To confirm this, the region around 1375 cm^{-1} should be examined for the presence of the umbrella mode band. There is a band at 1375 cm^{-1}, confirming we do have a methyl group. The band at 1397 cm^{-1} looks like it might be part of a split umbrella mode. However, as discussed in Chapter 2, umbrella mode splitting always occurs with the lower wavenumber band being more intense. The opposite is true in this spectrum, so the band at 1397 cm^{-1} is most likely a ring mode, and the saturated moieties are unbranched methyl groups.

The band at 1679 cm^{-1}, from its intensity and position, is a C=O stretch. This band could be the C=O stretch of an aromatic ketone, aromatic acid, or an amide (see Chapter 5). Examination of the rest of the spectrum shows an aromatic C-C-C stretch at 1250 cm^{-1}, confirming this is an aromatic ketone and also confirming the molecule contains a benzene ring.

The next question to answer is the substitution pattern on the benzene ring. There is no 690 cm^{-1} band, so the molecule must be either ortho- or para-substituted. The band at 829 cm^{-1} is assigned as the out-of-plane C-H bending vibration, and is in the correct range to make this a para- substituted molecule.

At this point, the molecular pieces we have are a para-substituted benzene ring, an aromatic ketone, and an individual methyl group(s). By definition, an aromatic ketone has a benzene ring attached to the carbonyl carbon, so one of the benzene ring substituents must be the C=O group. The second substituent on the carbonyl carbon could be a methyl group or another benzene ring, since we cannot easily distinguish between alkyl/aryl and diaryl ketones. The substituent on the benzene ring para to the ketone could also be a methyl group or another ketone. The small number of aromatic C-H stretches and ring modes argues for there being only one benzene ring. The simplicity of the spectrum argues for the benzene ring being symmetrically substituted. The only molecule that meets these conditions is p-diacetyl benzene.

Problem 4.3

Benzaldehyde (C_7H_6O)

Peak	Assignment
3087, 3065	Aromatic C-H stretch
2820, 2739	Aldehydic C-H stretch
1703	C=O Stretch
1598, 1585	Aromatic ring modes
1392	Aldehyde C-H bend
1204	Aldehyde C-C stretch
746	Aromatic out-of-plane C-H bend
688	Aromatic ring bend

The C-H stretches in this spectrum above 3000 cm^{-1} indicate unsaturation. The lack of a band between 1680 and 1630 cm^{-1}, and the ring modes around 1600 cm^{-1} confirm the unsaturation is an aromatic ring. There are two C-H stretches below 3000 cm^{-1}, but they fall well below the positions expected for the C-H stretches of methyl or methylene groups. These low wavenumber C-H stretches are characteristic of aldehydes, and the presence of an intense C=O stretch at 1703 cm^{-1} confirms that this is an aldehyde. The position of the C=O stretch also indicates this is an aromatic aldehyde, and that the benzene ring is directly attached to the carbonyl carbon. As confirmatory evidence that this is an aldehyde the aldehydic C-H stretch is found at 1392 cm^{-1}, and the C-C stretch involving the benzene ring/carbonyl carbon bond is at 1204 cm^{-1} with medium intensity.

Since we have a benzene ring, we need to determine its substitution pattern. There is a feature at 688 cm^{-1} that is assigned as the aromatic ring-bending band and means the molecule is either mono- or meta- substituted. The out-of-plane C-H bend at 746 cm^{-1} means the benzene ring is singly substituted. The molecule must be benzaldehyde.

Problem 4.4

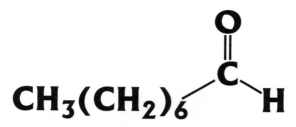

Octylaldehyde ($C_8H_{16}O$)

Peak	Assignment
2955	Methyl asymmetric C-H stretch
2928	Methylene asymmetric C-H stretch
2858	Methylene symmetric C-H stretch
2715	Aldehydic C-H stretch
1727	C=O Stretch
1393	Aldehydic C-H bend
1380	CH_3 Umbrella mode
724	CH_2 Rock

The group of bands just below 3000 cm^{-1} means this molecule is totally saturated. Methyl and methylene stretching bands are clearly seen. The methylene asymmetric stretch at 2928 cm^{-1} is more intense than the methyl asymmetric stretch at 2955 cm^{-1}, indicating there is a medium or long hydrocarbon chain present. To establish the number of methylenes in a row, a search for the CH_2 rock at 720 cm^{-1} must be made. It is present at 724 cm^{-1}, so the hydrocarbon chain in this molecule has at least 4 methylenes in a row.

There is a medium intensity but low wavenumber C-H stretch at 2715 cm^{-1}. This low wavenumber C-H stretch is indicative of an aldehyde, and the intense C=O stretch at 1727 confirms that this is a saturated aldehyde. We next need to determine if the alpha carbon of the aldehyde is branched or straight chain by looking for a band at either 2715-2700 cm^{-1} or 2730-2715 cm^{-1}, respectively. Unfortunately, the aldehydic C-H stretch falls at exactly 2715 cm^{-1}, so it is ambiguous as to whether or not the alpha carbon is branched. The bands at 1393 and 1380 cm^{-1} could be interpreted as a split umbrella mode. However, 1393cm^{-1} is in perfect position to be the aldehydic C-H stretch, so the band at 1380 cm^{-1} is the methyl umbrella mode, and the hydrocarbon chain is probably not branched. We have narrowed this interpretation down to a saturated straight chain aldehyde with four or more methylenes. Unfortunately, we cannot determine the exact number of methylenes from the spectrum. The molecule is octylaldehyde.

Problem 4.5

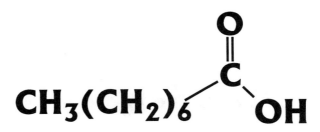

Octanoic Acid ($C_8H_{16}O_2$)

Peak	Assignment
3500-2500	O-H Stretch
2958	Methyl asymmetric C-H stretch
2929	Methylene asymmetric C-H stretch
2870	Methyl symmetric C-H stretch
2858	Methylene symmetric C-H stretch
1711	C=O Stretch
1414	In-plane O-H bend
1379	CH_3 Umbrella mode
1282	C-O Stretch
938	Out-of-plane O-H bend
725	CH_2 Rock

The broad feature between 3500 and 2500 cm^{-1} is undoubtedly the O-H stretch of a carboxylic acid. No other molecule has a feature this broad and intense. Superimposed on the O-H stretch is a series of four saturated C-H stretches. These are the methyl and methylene asymmetric and symmetric stretches. The relative intensity of the asymmetric stretches at 2958 and 2929 cm^{-1} indicates this is a medium length hydrocarbon chain. The CH_2 rocking band at 725 cm^{-1} means there are at least 4 methylenes in a row in the hydrocarbon chain. The methyl umbrella mode at 1379 cm^{-1} is not split, indicating the hydrocarbon is probably a straight chain.

To confirm that this molecule is a carboxylic acid, several other bands need to be tracked down. The carbonyl stretch is at 1711 and the C-O stretch is at 1282 cm^{-1}. The O-H in-plane and out-of-plane bends are at 1414 and 938 cm^{-1}. Unfortunately, we cannot determine the exact length of the hydrocarbon chain in this molecule, and are left with it being a saturated, straight chain, and medium length carboxylic acid. The molecule is octanoic acid.

Problem 4.6

Sodium Benzoate ($C_7H_5O_2Na$)

Peak	Assignment
3091-3028	Aromatic C-H Stretches
1597	Aromatic ring mode
1550	Asymmetric CO_2 stretch
1421	Symmetric CO_2 stretch
709	Aromatic out-of-plane C-H bend
682	Aromatic ring bend

There are C-H stretches above 3000 cm^{-1} in this spectrum but none below 3000 cm^{-1}, which means the molecule totally unsaturated. The lack of any bands between 1680 and 1630 cm^{-1}, and the strong ring mode at 1597 cm^{-1} confirms the unsaturation is an aromatic ring. To determine the substitution pattern on the benzene ring, the out-of-plane C-H bends at low wavenumber must be found. There are bands at 708 and 682 cm^{-1}, which are the out-of-plane C-H bend and the aromatic ring-bending mode, respectively. These two bands together indicate this is a monosubstituted benzene ring.

The nature of the substituent on the benzene ring can be ascertained from the two intense bands at 1550 and 1421 cm^{-1}. These are the asymmetric and symmetric CO_2 stretches of a carboxylate. Since the CO_2 group is attached to a benzene ring, and it is the only substituent, this carboxylate must be a salt of benzoic acid, referred to as a benzoate. Unfortunately, this spectrum contains no information as to the identity of the metal atom in the carboxylate. The molecule is sodium benzoate.

Problem 4.7

Benzyl Acetate ($C_9H_{10}O_2$)

Peak	Assignment
3067, 3036	Aromatic C-H stretches
2958, 2894	Methyl C-H stretches
1743	C=O Stretch
1382	CH_3 Umbrella mode
1230	Acetate ester C-C-O stretch
1028	Ester O-C-C stretch
751	Aromatic out-of-plane C-H bend
699	Aromatic ring bend

The C-H stretches above and below 3000 cm^{-1} indicate this molecule has saturated and unsaturated portions. The lack of bands between 1680 and 1630 cm^{-1} means this is not an alkene, but probably an aromatic ring. A search for ring modes turns up some weak, narrow bands between 1500 and 1350 cm^{-1} that might fit the bill. However, the best indicator for the presence of a benzene ring in this molecule is the out-of-plane C-H bend at 751 cm^{-1} and the ring-bending mode at 699 cm^{-1}. These two bands also establish the benzene ring as being mono-substituted. The two saturated C-H stretches are in about the right place to be due to a methyl group, and the umbrella mode at 1382 cm^{-1} confirms its presence. The presence of only two saturated C-H stretches argues against there being only one type of saturated carbon.

The most intense band in the spectrum at 1743 cm^{-1} is a C=O stretch. When combined with the next two most intense bands at 1230 and 1028 cm^{-1}, we find that this spectrum follows the rule of three, and that we have a saturated ester. Recall from Chapter 4 that acetate esters have a unique C-C-O stretching vibration at about 1240 cm^{-1}. The C-C-O stretch in this spectrum is at 1230 cm^{-1}, too high to be a generic saturated ester, but just about right to be an acetate ester. We can therefore conclude that this compound is an ester of acetic acid. By definition, the acetate group has a methyl group attached to the carbonyl carbon. This is where we would place the methyl group whose presence was discussed above.

The jigsaw puzzle pieces we have at this point are an acetate ester and a monosubstituted benzene ring. The benzene ring could be attached to the ester oxygen and be consistent with everything we know so far. However, the true structure of this molecule is benzyl acetate. The benzene ring is separated from the ester oxygen by a methylene group. Unfortunately, there is nothing in the spectrum that would tip us off that there was a methylene group present in this molecule.

Problem 4.8

Diisobutyl phthalate ($C_{16}H_{22}O_4$)

Peak	Assignment
3070	Aromatic C-H stretch
2962, 2874	Methyl C-H stretches
1728	C=O Stretch
1600, 1580	Aromatic ring modes
1393, 1376	Split CH_3 umbrella mode, intensity ratio 1:2
1286	Aromatic ester C-C-O stretch
1122	Aromatic ester O-C-C stretch
744	Aromatic out-of-plane C-H bend

The lone peak at 3070 cm^{-1} is suggestive of an unsaturation, but it is rather weak. Examination of the 1680 to 1630 cm^{-1} region shows no bands, so it is not an alkene. Bands at 1600 and 1580 cm^{-1} look like ring modes, and the strong out-of-plane C-H bend at 744 cm^{-1} confirms that the unsaturation is an aromatic ring. To determine the substitution pattern on the benzene ring, we need to check for the presence or absence of the ring-bending mode near 690 cm^{-1}. There are small features in this spectrum at 705 and 696 cm^{-1}, but they are too weak to be ring-bending bands. Thus, there is no ring-bending band, and the molecule is either ortho or para substituted. The out-of-plane C-H bend is in the right place for the benzene ring to be ortho-substituted.

The two C-H stretches below 3000 cm^{-1} are in perfect position to be methyl asymmetric and symmetric C-H stretches. A search for the methyl group umbrella mode turns up a split umbrella mode, with peaks at 1393 and 1376 cm^{-1}. This split umbrella mode means there is a branch point in the molecule. Note that the band at 1376 cm^{-1} is about twice the intensity of the band at 1393 cm^{-1}. This means the branch point is either an isobutyl or t-butyl group.

The three most intense bands in the spectrum are at 1728, 1286, and 1122 cm^{-1}. These bands follow the rule of three, which means this spectrum is that of an ester. The positions of these bands are correct to be the C=O, C-C-O, and O-C-C stretch of an aromatic ester. This means one of the substituents on the ortho-substituted benzene ring is an ester, and that the benzene ring is directly attached to the carbonyl carbon.

We have identified all the major peaks in the spectrum. The pieces of molecule we have at this point are a branched alkane, an ortho substituted benzene ring, and an aromatic ester. The alkane moiety could attach to either the benzene ring or the carbonyl carbon. However, this would leave a bond dangling, and there is no evidence of other substituents in the spectrum. The easiest way to solve this dilemma is to put the alkane portion on the carbonyl carbon and assume the two substituents on the benzene ring are the same. This assumption is correct and the molecule is diisobutyl phthalate. Note that the branch point in this molecule is the isobutyl group. It was not possible from this spectrum to distinguish between an isobutyl or t-butyl branch point. Phthalate esters like this one are commonly used as plasticizers in polymers.

Problem 4.9

Acetyl Salicylic Acid or Aspirin ($C_9H_8O_4$)

Peak	Assignment
3500-2500	Acid O-H stretch
3064	Aromatic C-H stretch
1753	C=O stretch
1691	C=O Stretch
1604, 1575	Aromatic ring modes
1418	Acid in-plane O-H bend
1369	Methyl umbrella mode
1305	Acid C-O stretch
1188	Saturated ester C-C-O
1094	Saturated ester O-C-C
917	Acid out-of-plane O-H bend
755	Aromatic out-of-plane C-H bend

This problem is probably the toughest in the book. However, by using the systematic approach to interpretation discussed in Chapter 1, and all the other information you have gathered so far in this book, the interpretation of this spectrum is doable.

The first thing to note about this spectrum is that there is a broad, strong feature at high wavenumber, and many sharp bands at low wavenumber. The broad feature is undoubtedly the O-H stretch of a carboxylic acid. Tracking down the other acid bands will be an easy task because of their widths. Acid carbonyl stretches occur from either 1730 to 1700 cm^{-1} for saturated acids and 1710 to 1680 cm^{-1} for aromatic acids. There is no band between 1700 and 1730 cm^{-1}, but there is a strong band at 1691 cm^{-1}; this is the C=O stretch of an aromatic acid. The acid in-plane and out-of-plane O-H bending bands are at 1418 cm^{-1} and 917 cm^{-1}, respectively. Last, the acid C-O stretch is at 1305 cm^{-1}.

Close examination of this spectrum will reveal not one but two carbonyl stretching bands. The second band, at 1753 cm^{-1}, is in range to be either a saturated ester or a saturated carbonate. Saturated carbonates have a C-O stretch from 1280-1240 cm^{-1}, but there are no bands in this region. However, there are bands at 1188 and 1094 cm^{-1}, the C-C-O and O-C-C stretches of an ester. All of the rule of three bands for our ester fall in the right range to be a saturated ester. Now, to have a saturated ester, there must be a saturated carbon attached to the carbonyl carbon. The intense O-H stretch masks any possible saturated C-H stretching vibrations. However, there is an umbrella mode at 1369 cm^{-1}, so there is a methyl group in this molecule.

The next problem to tackle is the substitution pattern on the benzene ring. There are several bands between 700 and 840 cm^{-1} that could be assigned as out-of-plane C-H bending vibrations. Typically, the most intense of these is the best one to use to determine ring substitution. It is difficult to see, but by holding this spectrum upside down, you can see that the most intense of these bands is at 755 cm^{-1}. The next thing to look for is the aromatic ring-bending mode at 690±10 cm^{-1}. There are no bands in this region, so the molecule must be either ortho or para substituted. The out-of-plane C-H bend at 755 cm^{-1} indicates this is an ortho-substituted molecule. Additionally, the fact that there is an aromatic acid and a saturated ester in this molecule means it must be disubstituted, at a minimum.

The jigsaw puzzle pieces we have are:

1. An ortho-substituted benzene ring
2. An aromatic acid
3. A saturated ester
4. A methyl group

The acid must be attached to the benzene ring since it is aromatic. We must next piece together the methyl group and the saturated ester in a way to make a single substituent that can attach to the benzene ring. Since the ester is saturated, the benzene ring is not attached to the carbonyl carbon. Since the methyl group is saturated, it must be attached to the carbonyl carbon giving the following moiety:

$$\text{O-C(=O)-CH}_3$$

This ester must be attached to the benzene ring via the ester oxygen or else it would not be a saturated ester. Therefore, the molecule is acetyl salicylic acid, the main ingredient in aspirin.

Problem 5.1

N-Ethylformamide (C₃H₇NO)

[Structure: H-C(=O)-N(H)-CH₂CH₃]

Peak	Assignment
3290	N-H Stretch
3058	N-H Bend overtone
2979, 2878	Methyl C-H stretches
2939	CH₂ Asymmetric stretch
1666	C=O Stretch
1538	In-plane N-H bend
1384	CH₃ Umbrella mode
1243	Amide C-N stretch
728	N-H Out-of-plane bend

The medium intensity, medium width band at 3290 cm^{-1} is an N-H stretch. This band is too weak and narrow to be an O-H stretch. The fact that there is just one N-H stretch means this molecule has just one N-H bond, and thus contains a secondary nitrogen. The band at 3058 cm^{-1} is in the right place to be an unsaturated C-H stretch. However, it is much too broad to be a C-H stretch. This band is actually an overtone of a fundamental band that appears at 1538 cm^{-1}.

The three bands just below 3000 cm^{-1} are saturated C-H stretches. The bands at 2979 and 2878 cm^{-1} are the asymmetric and symmetric stretches of a methyl group. It is difficult to see, but there is a weak band at 2939 cm^{-1}, which is the asymmetric C-H stretch of a methylene group. The methylene symmetric stretch is too weak to be seen. Note how much more intense the CH₃ bands are than the CH₂ band. This indicates the saturated hydrocarbon chain in this molecule is a short one.

To confirm the assignment of a methyl group, we must search for the methyl umbrella mode. It is found in this spectrum at 1384 cm^{-1}, and is a single band indicating the hydrocarbon chain is not branched. To establish the number of methylenes in a row, we must find the CH₂ rock near 720 cm^{-1}. There is a band at 728 cm^{-1}, but it is much too broad to be a C-H bending band. Therefore, we must conclude that there are less than 4 CH2s in a row in the hydrocarbon chain. This is consistent with the low intensity of the CH₂ stretching band.

The next band in the spectrum with any intensity is at 1666 cm^{-1} and, by its position and intensity, it must be a C=O stretch. The fact that the molecule also contains a secondary nitrogen atom is highly suggestive of a secondary amide. The band next to the C=O stretch at 1538 cm^{-1} is an N-H in-plane bend, and is diagnostic for secondary amides. Other bands from the secondary amide must now be found. The C-N stretch is at 1243 cm^{-1} and the broad, out-of-plane N-H bend is at 728 cm^{-1}.

The molecular fragments we have at this point are a short hydrocarbon chain and a secondary amide. The hydrocarbon chain must be attached to the amide nitrogen to make it a secondary amide. However, we do not know what substituents, if any, are attached to the carbonyl carbon. The molecule is N-ethylformamide. Note that the other substituent on the carbonyl carbon is a hydrogen atom.

Problem 5.2

Kevlar® Polyamide (C₇H₅NO)ₙ
Polyparaphenylene terephthalamide

Peak	Assignment
3320	N-H Stretch
3059	Aromatic C-H stretch
1660	C=O Stretch
1543	In-plane N-H bend
1516, 1404	Aromatic ring modes
1262	Amide C-N stretch
824	Aromatic out-of-plane bend C-H bend

The medium intensity, medium width band at 3320 cm^{-1} is an N-H stretch. Since there is only one of these bands, there must be a secondary nitrogen atom in the molecule. There are no C-H stretches below 3000 cm^{-1}, so this molecule is totally unsaturated. The small bump at 3059 cm^{-1} is probably an unsaturated C-H stretch. If there were an alkene present, we would look for a narrow, medium intensity band between 1680 and 1630 cm^{-1}. What we find is a strong, intense band at 1660 cm^{-1}. Although it is in the right place, this band does not have the right intensity or width to be a C=C stretch. Therefore, this is not an alkene. Examination of the 1600 to 1400 cm^{-1} region shows some bands that might be ring modes, and the aromatic out-of-plane C-H bending band at 824 cm^{-1} confirms we have an aromatic ring. There is no ring-bending band at 690 cm^{-1}, so the molecule must be ortho or para (and at a minimum is disubstituted). Based on the out-of-plane C-H bend, this molecule is para substituted.

The band at 1660 cm^{-1} is strong and is in the right place to be a C=O stretch. The fact that we also have a secondary nitrogen in the molecule is highly suggestive of a secondary amide. This is confirmed by the presence of a secondary amide N-H bending band at 1543 cm^{-1}. The C-N stretch of the amide is found at 1262 cm^{-1}. The N-H wag (out-of-plane bend) is typically seen from 750 to 680 cm^{-1}. However, this spectrum cuts off at 750 cm^{-1} because of the MCT detector on the infrared microscope used to obtain the spectrum.

Thus, the molecular pieces we have at this point are a para-substituted benzene ring and a secondary amide. The lack of bands due to other functional groups argues for the ring being symmetrically substituted. The fact that this sample is a fiber indicates it is made from a polymer. By putting the amide groups at both ends of the benzene ring, we come up with the molecule. It is Kevlar® polyamide fiber (Polyparaphenylene terephthalamide) made by Dupont.

Problem 5.3

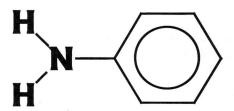

Aniline (C$_6$H$_7$N)

Peak	Assignment
3436	Asymmetric NH$_2$ Stretch
3358	Symmetric NH$_2$ Stretch
3216	NH$_2$ Bend overtone
3072, 3078	Aromatic C-H stretches
1622	NH$_2$ Scissors
1603, 1499	Aromatic ring modes
1278	Aromatic amine C-N stretch
754	Aromatic out-of-plane bend C-H bend
693	Aromatic ring bend
850-750 (broad)	NH$_2$ Out-of-plane bend

The two high wavenumber, medium width, medium intensity bands at 3436 and 3358 cm^{-1} are the asymmetric and symmetric stretches of an NH$_2$ group. Additionally, both these bands are perfectly positioned to be due to an aromatic amine. Since there are two N-H stretching bands, this molecule must be a primary aromatic amine. These bands are at too high a wavenumber to be from an amide. The band at 3216 cm^{-1} is in range to be an unsaturated C-H stretch, but it is much too broad to qualify for that assignment.

The bands at 3072 and 3038 cm^{-1} are unsaturated C-H stretches. There are no saturated C-H stretches below 3000 cm^{-1}, so this molecule is totally unsaturated. To determine the type of unsaturation, we need to look to lower wavenumber. The band at 1499 cm^{-1} looks like an aromatic ring mode. The bands at 754 and 693 cm^{-1} can be assigned as the out-of-plane C-H bend and ring bend of a benzene ring. These bands confirm that the source of unsaturation is a benzene ring. The band at 693 cm^{-1} and the position of the C-H bending band mean this is a mono-substituted benzene ring. Since we know we have a primary aromatic amine and a monosubstituted benzene ring, putting the two together means the molecule must be aniline.

Although the two strong bands at 1622 and 1603 cm^{-1} are reminiscent of a carbonyl stretch, they are a little too weak, narrow, and too low in wavenumber to be a carbonyl stretch. Recalling aniline is a primary aromatic amine, the bands at 1622 cm^{-1} are correctly assigned as the NH$_2$ scissors and an aromatic ring mode. The amine C-N stretch is at 1278 cm^{-1}, again confirming that the NH$_2$ is attached to the aromatic ring. The NH$_2$ wag is a broad feature around 800 cm^{-1} with the narrower benzene ring bands at 754 and 693 cm^{-1} superimposed upon it.

Problem 5.4

$$CH_3CH_2CH_2CH_2-N\begin{matrix}H\\ \\CH_3\end{matrix}$$

N-methylbutylamine ($C_5H_{13}N$)

Peak	Assignment
3294	N-H Stretch
2961, 2872	CH_3 Asymmetric and symmetric stretches
2872	CH_2 asymmetric stretch
2793	N-CH_3 Symmetric stretch
1378	CH_3 Umbrella mode
1151, 1127	C-N Stretches
739	N-H Wag

The band at 3294 cm^{-1} is an N-H stretch (it is too small and too narrow to be an O-H stretch). Since it is by itself, the sample contains a secondary nitrogen. There are no C-H stretches above 3000 cm^{-1}, but a series of them below 3000 cm^{-1}. This means the molecule is totally saturated. The bands at 2961 and 2872 are in range to be the asymmetric and symmetric stretches of the CH_3 group. The band at 2931 cm^{-1} is the CH_2 asymmetric stretch. The symmetric stretch of this functional group is not seen, probably because it is too weak to compete with the CH_3 stretching bands. This fact, combined with the equal intensity of the methyl and methylene asymmetric stretches, indicates the presence of a short hydrocarbon chain.

To determine whether the hydrocarbon chain is branched, the CH_3 umbrella mode must be found. It is clearly seen at 1378 cm^{-1} and is a single peak. This means the hydrocarbon chain is not branched. To determine the number of methylenes in a row in the hydrocarbon chain, the CH_2 rocking band near 720 cm^{-1} must be found. There is a band in this spectrum at 739 cm^{-1}, but it is much too broad to be from a C-H bending vibration. If the CH_2 rocking band were present, it would probably appear as a sharp spike superimposed upon the broad band at 739 cm^{-1}. Therefore, the hydrocarbon chain must have less than 4 methylenes in a row and must be relatively short. This is consistent with other evidence discussed above.

The secondary nitrogen is not part of an amide, since there is no C=O stretch present in this spectrum. If it were part of a secondary amine, it should have an N-H wag band between 750 and 700 cm^{-1}. The broad band at 739 cm^{-1} fits the bill perfectly, confirming that this molecule is a secondary amine. The C-N stretch of this amine is probably the two bands at 1151 or 1127 cm^{-1}.

We know we have a short hydrocarbon chain attached to the amine nitrogen, but what is the second hydrocarbon substituent on the nitrogen? Look closely at the C-H stretches. There is a band at 2793 cm^{-1}. It is much too low to be due to a normal CH_2 or CH_3. However, the symmetric stretch of a CH_3 attached to a nitrogen can have a band in exactly this wavenumber range. Thus, the second substituent on the nitrogen is a methyl group.

So, we have a secondary amine whose substituents are a methyl group and a short hydrocarbon chain. The only thing left to determine is the exact number of methylene groups in the molecule, which is not possible from the infrared spectrum alone. The molecule is N-methylbutylamine.

Problem 5.5

Benzonitrile (C$_7$H$_5$N)

Peak	Assignment
3065	Aromatic C-H Stretch
2228	Aromatic C≡N Stretch
1598, 1490, 1447	Aromatic ring modes
758	Out-of-plane C-H bend
687	Aromatic ring bend

The only band near 3000 cm^{-1} is at 3065 cm^{-1}, and is in position to be an unsaturated C-H stretch. Note that this molecule is totally unsaturated. There are no bands between 1680 and 1630 cm^{-1}, so the unsaturation is not a C=C bond. A look at the rest of the spectrum shows some intense, narrow bands between 1600 and 1450 cm^{-1} that could be aromatic ring modes. The two most intense bands in the spectrum are at 758 and 687 cm^{-1}, and are easily assigned as aromatic ring out-of-plane C-H bending and ring-bending bands. These bands confirm that the unsaturation is a benzene ring.

Because of the ring-bending band, the benzene ring substitution pattern must be either mono or meta. Normally, the out-of-plane C-H bend would be used to distinguish between these. Unfortunately, this band falls at 758 cm^{-1}, which can be either mono or meta. So, the ring substitution pattern cannot be known.

The last intense band in the spectrum left to assign is at 2228 cm^{-1}, and is the C≡N stretch of an aromatic nitrile. The position of this band tells us that there is a nitrile on the benzene ring. There are no other major bands in the spectrum and the simplicity of the spectrum argues for mono substitution. The molecule is benzonitrile.

Problem 5.6

O-Nitrotoluene ($C_7H_7NO_2$)

Peak	Assignment
3070	Aromatic C-H Stretch
2976, 2862	CH_3 Stretches
2862	NO_2 Stretch overtone?
1612	Aromatic ring mode
1524	NO_2 Asymmetric stretch
1382	CH_3 Umbrella mode
1348	NO_2 Symmetric stretch
858	NO_2 Scissors
788, 730	Aromatic out-of-plane C-H bends

There are C-H stretches above and below 3000 cm^{-1} in this spectrum, indicating it contains saturated and unsaturated moieties. The C-H stretch above 3000 cm^{-1}, the sharp bands around 1600 and 1500, and the medium intensity bands around 750 cm^{-1} all confirm that the source of unsaturation is a benzene ring. There are three saturated C-H stretches. The bands at 2976 and 2862 cm^{-1} are in position to be methyl asymmetric and symmetric stretches. The band at 2934 cm^{-1} could be assigned as a methylene asymmetric stretch (more on this later). The search for the CH_3 umbrella mode turns up a band at 1382 cm^{-1}, confirming there is a methyl group, and that there is no hydrocarbon chain branching. To determine how many, if any, methylene groups there are in a row, one would look for the CH_2 rock at 720 cm^{-1}. There is a band at 730 cm^{-1}, but it is much too intense to be due to a CH_2 group (its an aromatic out-of-plane C-H bend). If there were a methylene rock band present, it would be obscured by the aromatic band.

The overriding spectral feature of this spectrum are the two intense bands at 1524 and 1348 cm^{-1}, which are easily identified as the asymmetric and symmetric NO_2 stretches of the nitro group. The scissors band of the nitro group is at 858 cm^{-1}. Since the vast majority of nitro compounds in the world are aromatic and we have an aromatic ring in our molecule, it is safe to assume that the nitro group is attached to the benzene ring. Recall, however, that nitro groups make it difficult to determine benzene ring substitution patterns. Thus, we have an aromatic nitro compound, a methyl group, and maybe a methylene. There is not much more that can be learned from this spectrum. In reality, the molecule is ortho-nitrotoluene. The only way to determine the benzene ring substitution pattern in this case is by some other means, such as NMR spectroscopy. The band at 2934 cm^{-1} is not due to a methylene group, but is possibly an overtone of the NO_2 asymmetric stretch. The fact that the 2934 cm^{-1} band is sharper than the methyl stretches around it may tip one off that it is not a C-H stretch, but this is by no means obvious, even to a trained observer.

Appendix 2

Group Wavenumber Tables

You will find listed here all the group wavenumber tables contained in the body of this book. These tables are the best summaries of the information contained herein. You do not need to commit these numbers to memory, but keep these tables handy whenever you need to interpret a spectrum. All numbers in the tables are in cm^{-1}.

Table 1.1 Functional Groups Easily Spotted in a Spectrum

Band Position in cm^{-1}	**Functional Group**
3500-3200	O-H or N-H
3200-2800	C-H
2250-2000	C≡N, C≡C
1800-1600	C=O
<1000	C=C, Benzene Rings

Table 2.1. The Wavenumber Ranges for the C-H Stretches of Alkanes (in cm^{-1})

Vibration	**Wavenumber Range**
CH$_3$ Asymmetric	2962±10
CH$_3$ Symmetric	2872±10
CH$_2$ Asymmetric	2926±10
CH$_2$ Symmetric	2855±10

Table 2.2. The C-H Bending Vibrations of Alkanes (all numbers in cm^{-1})

Vibration	**Wavenumber Range**
CH$_3$ Asymmetric bend	1460± 10
CH$_3$ Symmetric bend (umbrella mode)	1375±10
CH$_2$ Scissors	1455±10
CH$_2$ Rock	720±10

Table 2.3 C-H Bending and Stretching Bands for Branched Alkanes (all numbers in cm^{-1})

Vibration	**Wavenumber**
Methine C-H Stretch	~2900 (weak)
Split umbrella mode of isopropyl and *gem*-dimethyl groups	1385 - 1365 (2 bands) Intensity ratio ~1:1
Split umbrella mode of isobutyl and *t*-butyl groups	1395 - 1365 (2 bands) Intensity ratio ~1:2
Methine C-H bend	~1350 (weak)

Table 2.4. Positions of C-H Stretches for Saturated and Unsaturated Functional Groups (all numbers in cm^{-1})

Functional Groups	C-H Stretch Position
Saturated (CH$_3$, CH$_2$, CH)	< 3000 cm^{-1}
Unsaturated (C=C, C≡C, Aromatic Rings)	> 3000 cm^{-1}
Vinyl/vinylidine CH$_2$ asymmetric stretch	3090-3075

Table 2.5 The C=C Stretching, C-H Stretching, and C-H Bending Bands of Alkenes

Substitution Pattern	C-H Stretch	C=C Stretch	Out-of-Plane C-H Bend
Vinyl	3090-3075	1660-1630	990±5, 910±5
Vinylidine	3090-3075	1660-1630	890±5
Cis	3050-3000	1660-1630	690±50
Trans	3050-3000	1680-1665	965±5
Trisubstituted	3050-3000	1680-1665	815±25
Tetrasubstituted	-	1680-1665	-

Table 2.6 Summary of Group Wavenumbers for Alkynes (all numbers in cm^{-1})

Substitution	C≡C Stretch	C-H Stretch	C-H Wag
Monosubstituted	2140-2100	3350-3250	700-600
Disubstituted	2260-2190	-	-

Table 2.7 Bands for Mono- and Disubstituted Benzene Rings (all numbers in cm^{-1})

Substitution Pattern	Out-of-Plane C-H Bending	Ring Bend (690±10 cm^{-1})
Mono	770-710	Yes
Ortho	810-750	No
Meta	770-735	Yes
Para	860-790	No

Table 2.8 Bands for Methyl Groups Bonded to Benzene Rings (all numbers in cm^{-1})

Vibration	Wavenumber
CH$_3$ Symmetric stretch	2925±5
CH$_3$ Bend overtone	2865±5

Table 3.1 The Diagnostic Infrared Bands for Alcohols (all numbers in cm^{-1})

Subst. Pattern	C-O Stretch	O-H Stretch	O-H Bends
All	-	3350±50	1350±50, 650±50
1°	1075-1000	"	"
2°	1150-1075	"	"
3°	1210-1100	"	"
Phenols	1260-1200	"	"

Table 3.2 Group Wavenumbers for Ethers (all numbers in cm^{-1})

Ether Type	Asymmetric C-O-C	Symmetric C-O-C
Saturated, Unbranched	1140-1070 (1 band)	890-820
Saturated, Branched	1210-1070 (2 or more bands)	890-820
Alkyl/Aryl (mixed)	1300-1200 and 1050-1010	-
Aryl	1300-1200	-

Table 3.3 The C-H Stretches and Bends for Hydrocarbons Attached to Oxygen (all numbers in cm^{-1})

Vibration	CH$_3$-O	CH$_2$-O
Asymmetric C-H stretch	2970-2920	2955-2920
Symmetric C-H stretch	2830±10	2878-2835
CH$_3$ Umbrella mode	1470-1440	-

Table 4.1 The Group Wavenumbers of Ketones (all numbers in cm^{-1})

Vibration	Wavenumber Range
Saturated C=O stretch	1715±10
Aromatic C=O stretch	1700-1640
Saturated C-C-C stretch	1230-1100
Aryl/Alkyl and Diaryl C-C-C stretch	1300-1230

Table 4.2 The Group Wavenumbers of the Aldehyde Functional Group (in cm^{-1})

Vibration	Wavenumber Range
Saturated C=O Stretch	1730±10
Aromatic C=O Stretch	1710-1685
C-H Bend	1390±10
Aldehydic C-H Stretch, general	2850-2700 (1 or 2 bands)
Aldehydic C-H Stretch, unbranched α carbon	2730-2715 (1 band)
Aldehydic C-H Stretch, branched α carbon	2715-2700 (1 band)

Table 4.3 The Group Wavenumbers of Carboxylic Acids (all numbers in cm^{-1})

Vibration	Wavenumber Range
Saturated C=O Stretch	1730-1700
Aromatic C=O Stretch	1710-1680
C-O Stretch	1320-1210
O-H Stretch	3500-2500 (broad and intense)
O-H In-plane bend	1440-1395
O-H Out-of-plane bend	960-900

Table 4.4 The Group Wavenumbers of Carboxylates (all numbers in cm^{-1})

Vibration	Wavenumber Range
Asymmetric CO$_2$ stretch	1650-1540
Symmetric CO$_2$ stretch	1450-1360

Table 4.5 The Group Wavenumbers for Noncyclic Acid Anhydrides (all numbers in cm^{-1})

Vibration	Wavenumber Range
Saturated Symmetric C=O stretch	1820±5 (stronger)
Saturated Asymmetric C=O stretch	1750±5 (weaker)
Unsaturated Symmetric C=O stretch	1775±5 (stronger)
Unsaturated Asymmetric C=O stretch	1720±5 (weaker)
C-O Stretch	1060-1035

Table 4.6 The Group Wavenumbers for Cyclic Acid Anhydrides (all numbers in cm^{-1})

Vibration	Wavenumber Range
Saturated Symmetric C=O stretch	1870-1845 (weaker)
Saturated Asymmetric C=O stretch	1800-1775 (stronger)
Unsaturated Symmetric C=O stretch	1860-1840 (weaker)
Unsaturated Asymmetric C=O stretch	1760-1780 (stronger)
C-O and C-C Stretches	960-880, 1300-1000

Table 4.7 The Rule of Three Bands For Saturated Esters (all numbers in cm^{-1})

Vibration	Wavenumber Range
C=O Stretch	1750-1735
C-C-O Stretch	1210-1160
Acetate C-C-O Stretch	~1240
O-C-C Stretch	1100-1030

Table 4.8 The Rule of Three Bands For Aromatic Esters (all numbers in cm^{-1})

Vibration	Wavenumber Range
C=O Stretch	1730-1715
C-C-O Stretch	1310-1250
O-C-C Stretch	1130-1000

Table 4.9 Group Wavenumbers for Organic Carbonates (all numbers in cm^{-1})

Vibration	Saturated Carbonate	Mixed Carbonate	Aromatic Carbonate
C=O Stretch	1740±10	1790-1760	1820-1775
O-C-O Stretch	1280-1240	1250-1210	1220-1205

Table 5.1 The Group Wavenumbers of Primary Amides (all numbers in cm^{-1}).

Vibration	Wavenumber Range
NH$_2$ Stretches	3370-3170 (2 bands)
C=O Stretch	1680-1630
NH$_2$ Scissors	1650-1620
C-N Stretch	1430-1390
NH$_2$ Wag	750-600 (broad)

Table 5.2 The Group Wavenumbers of Secondary Amides (all numbers in cm^{-1}).

Vibration	Wavenumber Range
N-H Stretch	3370-3170 (1 band)
C=O Stretch	1680-1630
N-H In-plane bend	1570-1515 (strong)
C-N Stretch	1310-1230
N-H out-of-plane bend	750-680 (broad)

Table 5.3 The Group Wavenumbers of Tertiary Amides (all numbers in cm^{-1}).

Vibration	Wavenumber Range
C=O Stretch	1680-1630
C-N(CH$_3$)$_2$ Stretch	~1505

Table 5.4 The Group Wavenumbers of Imides (all numbers in cm^{-1}).

Vibration	Straight Chain	Cyclic
N-H Stretch	3200±50	3200±50
C=O Stretch(es)	1740-1670 (1 band)	1790-1735, 1750-1680 (2 bands)
N-H In-plane bend	~1505	-
C-N Stretch	1235-1165	-

Table 5.5 The Group Wavenumbers of the Primary Amine Group (all numbers in cm^{-1}).

Vibration	Saturated	Aromatic
NH$_2$ Asymmetric stretch	3380-3350	3500-3420
NH$_2$ Symmetric stretch	3310-3280	3420-3340
NH$_2$ Scissors	1650-1580	1650-1580
C-N Stretch	1250-1020	1350-1250
NH$_2$ Out-of-plane bend	850-750	850-750

Table 5.6 The Group Wavenumbers of the Secondary Amine Group (all numbers in cm^{-1}).

Vibration	Saturated	Aromatic
N-H Stretch	3320-3280	~3400
C-N Stretch	1180-1130	1350-1250
N-H Wag	750-700	750-700

Table 5.7 N-Methyl Symmetric Stretches (all numbers in cm^{-1}).

Grouping	Saturated	Aromatic
N-CH$_3$	2805-2780	2820-2810
N-(CH$_3$)$_2$	2825-2810, 2775-2765	2810-2790

Table 5.8 The C≡N Stretches of Nitriles (all numbers in cm^{-1}).

Vibration	Saturated	Aromatic
C≡N Stretch	2260-2240	2240-2220

Table 5.9 The Group Wavenumbers of the Nitro Functional Group (all numbers in cm^{-1}).

Vibration	Aromatic
NO$_2$ Asymmetric stretch	1550-1500
NO$_2$ Symmetric stretch	1390-1330
NO$_2$ Scissors	890-835

Table 6.1 The Group Wavenumber for Thiols (all numbers in cm^{-1}).

Vibration	Wavenumber Range
S-H Stretch	2590-2560

Table 6.2 The Group Wavenumbers for Compounds with a Single S=O Bond

Functional Group	S=O Stretch (cm^{-1})
Sulfoxide	1070-1030
Sulfite	1240-1180

Table 6.3 The Group Wavenumbers for SO$_2$ Containing Molecules (all numbers in cm^{-1})

Functional Group	Asymm. SO$_2$ Stretch	Sym. SO$_2$ Stretch
Sulfone	1340-1310	1165-1135
Sulfonate	1430-1330	1200-1150
Sulfate	1450-1350	1230-1150

Table 6.4 The Group Wavenumbers of Siloxanes (all numbers in cm^{-1})

Functional Group	CH$_3$ Umbrella Mode	CH$_3$ Rock	CH$_2$ Wag and Rock
Si-CH$_3$	1260± 5	-	-
Si(CH3)$_2$	1260± 5	800±10	-
Si(CH$_3$)$_3$	1260± 5	~840, ~760	-
Si-CH$_2$	-	-	1250-1200, 760-670

Table 6.5 The C-X Stretches for Halogenated Organic Molecules

Bond	C-X Stretch
C-F	1300-1000
C-Cl	800-600
C-Br	650-550
C-I	570-500

Table 7.1 The Group Wavenumbers of Inorganic Sulfates

Vibration	Wavenumber Range (cm^{-1})
S-O Stretch	1140-1080
S-O Bends	680-610

Table 7.2 The Group Wavenumbers of Silica

Vibration	Wavenumber Range (cm^{-1})
Silanol SiO-H Stretch	3700-3200
Si-O-Si Asymmetric stretch	1200-1000
Silanol Si-O Stretch	~940
Si-O-Si Symmetric stretch	~805
Si-O-Si Bend	~450

Table 7.3 The Group Wavenumbers for Inorganic Carbonates

Vibration	Wavenumber Range (cm^{-1})
C-O Stretch	1510-1410
C-O Out-of-plane bend	880-860
C-O In-plane bend	~740

Table 7.4 The Group Wavenumbers for Nitrates

Vibration	Wavenumber Range (cm^{-1})
N-O Stretch	1400-1340
N-O Out-of-plane bend	840-810
N-O In-plane bend	~720

Table 7.5 The Group Wavenumbers of Phosphates

Vibration	Wavenumber (cm^{-1})
PO_4^{-3} Stretch	1100-1000 (broad and strong)
PO_4^{-3} Bend	600-500

Table 8.1 Group Wavenumbers for Diisocyanates and Urethanes

Vibration	Wavenumber (cm^{-1})
Isocyanate asymmetric N=C=O stretch	2280-2240
Secondary Urethane N-H Stretch	3340-3250
Secondary Urethane C=O	1725-1705
Tertiary Urethane C=O	1690-1680
Secondary Urethane N-H Bend	1540-1520
Urethane C-O Stretch	~1250

Glossary

Many of the technical terms used in this book are defined here in the Glossary. Most words that appeared in italics in the body of the book have been included. Words that appear in italics in the Glossary are defined elsewhere in this section.

Acid Anhydride - A molecule formed by the reaction of two *carboxylic acid* molecules with each other. This reaction produces a water molecule as a byproduct. The acid anhydride functional group contains two *carbonyl* groups connected by an oxygen atom.

Alcohol - A functional group that contains the C-OH linkage.

Aldehyde - A functional group where a hydrogen atom and a carbon atom are attached to a *carbonyl carbon*.

Aliphatic Hydrocarbon - Any *hydrocarbon* molecule that does not contain an *aromatic* ring.

Alkanes - An important family of *hydrocarbon* molecules that contain strictly C-H and C-C bonds. Alkanes are an example of *saturated* molecules.

Alkenes - A type of *hydrocarbon* that contains at least one C=C bond. Alkenes are examples of *unsaturated* molecules.

Alkynes - A *hydrocarbon* molecule that contains at least one C≡C bond.

Alpha Carbon - A carbon atom bonded to a *carbonyl carbon*.

Amide - A functional group that contains a *carbonyl carbon* with a nitrogen atom and a carbon atom bonded to it. The nitrogen atom may have one or two hydrogen atoms attached to it.

Amine - A functional group that contains a nitrogen atom bonded to one, two, or three carbon atoms. The nitrogen atom may have one or two hydrogen atoms attached to it.

Anharmonic Motion - The opposite of *harmonic motion*. When applied to vibrating molecules, it means the atoms involved in a vibration do not vibrate in phase or with the same amplitude.

Aniline - An *amine* with an aromatic ring bonded to the nitrogen atom.

Aromatic - Describes a special type of carbon ring structure where the electrons in adjacent C-C bonds delocalize, spreading out the electron density around the ring. This gives the molecule a unique chemical stability and structure. The most commonly found aromatic moieties described in the book are aromatic *benzene* rings.

Beer's Law - An equation that relates the absorbance of light by a sample to the concentration, pathlength, and absorptivity of the sample. Typically used to measure the concentrations of molecules in unknown samples.

Benzene - An *aromatic hydrocarbon* molecule of chemical formula C_6H_6. Because of its *aromatic* properties, benzene has a unique structure. It is a regular hexagon with a carbon atom at each vertex, and one hydrogen attached to each carbon atom. It is the prototype for a large family of *aromatic* molecules.

Branched Alkane - Any *alkane* molecule that contains a *branch point*.

Branch Point - A point in a *hydrocarbon* chain where a carbon atom has three or more C-C bonds.

Carbonates - A functional group that contains the grouping CO_3. In organic carbonates, there are carbon atoms attached to two of the oxygens and the molecule has no charge. In inorganic carbonates, the CO_3 grouping has a -2 charge, and is ionically bonded to two positive ions, usually metal ions.

Carbonyl - A carbon/oxygen double bond, C=O.

Carbonyl Carbon - The carbon atom in a *carbonyl* group.

Carboxylate - A molecule made by the reaction of a *carboxylic acid* with a base. A carboxylate is a *salt*, and has the formula CO_2 with a charge of -1.

Carboxylic Acid - A functional group containing a COOH grouping. It consists of a *carbonyl* with a carbon atom and O-H group attached to the *carbonyl carbon*.

Chemical Environment - The immediate surroundings in which a molecule finds itself. The nature of the chemical environment around a molecule is determined by the type, number, and distance to nearest neighbor molecules.

Cis - an *isomer* of disubstituted *alkenes*, where the two carbon-bearing substituents are on the same side of the double bond.

Combination Band - Any band that appears in the infrared spectrum of a molecule due to excitation of two or more vibrations at the same time.

Conjugation - When the electron density in adjacent chemical bonds "spreads out" across the bonds. This often times leads to an *electronic effect*, which affects the spectrum of a molecule.

Diisocyanates - An isocyanate contains the grouping -N=C=O. A diisocyanate is any molecule that contains two of these groupings.

Dipole Moment - A measure of charge asymmetry, dipole moments are generated when two charges are held apart. The magnitude of a dipole moment is equal to the size of the charges times the distance between them.

Electric Vector - The electric part of light (*electromagnetic radiation*). The electric vector interacts with molecules to give rise to *infrared absorption*.

Electromagnetic Radiation - Another term for light. Light consists of two waves, an electric wave (the *electric vector*) and a magnetic wave that oscillate in planes perpendicular to each other, and perpendicular to the direction of travel of the light ray.

Electronegativity - A measure of how tightly an atom holds onto its bonding electrons. Fluorine is the most electronegative element, carbon is in the middle, and metals are the least electronegative elements.

Electronic Effects - Changes in the electronic structure of a molecule that cause infrared band positions to change.

Esters - A functional group that consists of a *carbonyl* group with a carbon atom and oxygen atom attached to the *carbonyl carbon*.

Ether - A functional group that consists of a central oxygen atom, with two carbon atoms singly bonded to it (C-O-C).

Fermi Resonance - A type of *vibrational interaction* involving a stretching and a bending vibration interacting with each other.

Fingerprint Region - The region between 1600 and 1000 cm^{-1} in a spectrum where many different bands appear. The appearance of this region is unique enough to each molecule to be considered a fingerprint of the molecule.

Fourier Transform Infrared Spectrometer (FTIR) - a type of *infrared spectrometer* that is capable of measuring an infrared spectrum quickly and with a good signal-to-noise ratio.

Frequency - A property of light that describes the number of cycles (undulations) that a light wave undergoes per second. Measured in *Hertz*, which have units of sec^{-1}.

Functional Groups - A bond or group of bonds that exhibit chemical and spectroscopic properties unique to that group.

Fundamental Transition - An infrared absorption in which a molecule is excited from the V=0 to the V=1 state (see *vibrational quantum number* for a definition of V). Most of the bands in *mid-infrared* spectra are due to fundamental transitions.

Gem-dimethyl - A *branch point* where two methyl groups are attached to a *quaternary carbon*.

Ground Vibrational State - When the *vibrational quantum number* of a molecule (V) equals zero. This corresponds to the lowest vibrational energy that a molecule can have.

Group Wavenumber - A diagnostic infrared band position for a *functional group*, measured in wavenumbers.

Harmonic Motion - When used to describe a vibrating chemical bond, motion where the two atoms in a bond vibrate in-phase with each other at the same amplitude.

Harmonic Oscillator - Any system that undergoes *harmonic motion*. Also, a simple model of molecular vibrations.

Hertz - The quantity used to denote the *frequency* of light, whose units are sec^{-1}.

Hit Quality Index (HQI) - A number, used in *library searching*, that describes how similar two spectra are to each other.

Hydrocarbon - A molecule that contains only carbon and hydrogen atoms.

Hydroxyl - A term for the O-H structural unit.

Hydroxyl Carbon - A carbon atom with a *hydroxyl* group attached to it.

Identities - The comparison of two spectra to each other to ascertain whether two samples are identical or not.

Imides - A functional group containing two *carbonyl* groups attached to a nitrogen atom.

Infrared Active - Molecular vibrations that can be excited by absorption of infrared radiation.

Infrared Inactive - Molecular vibrations that cannot be excited by absorption of infrared radiation.

Infrared Spectrometer - An instrument capable of measuring an *infrared spectrum*.

Infrared Spectroscopy - The study of the interaction of infrared light with matter.

Infrared Spectrum - A plot of measured infrared intensity versus some property of light, usually *wavenumber* or *wavelength*.

Isobutyl - A *branch point* consisting of an *isopropyl* group attached to a *methylene* group.

Isomers - Two molecules with the same chemical formula but different chemical structures.

Isopropyl - A *branch point* where two methyl groups are attached to a *methine* group.

Isotopes - Two atoms that have the same atomic number but different atomic masses.

Library Searching - An interpretation aid where an unknown spectrum is compared mathematically to a collection of known spectra. Spectra that are similar or identical are assumed to be from samples with similar or identical chemical compositions.

Mass Effect - Shifts in infrared band positions caused by changing the mass of the atoms involved in a vibration.

Mercaptans - See *thiols*.

Meta - An *isomer* of a disubstituted *benzene* ring, where the two substituents are separated by one carbon atom.

Methine - A functional group containing a carbon atom with one C-H bond and three C-C bonds.

Methoxy - The O-CH_3 structural unit.

Methyl - The CH_3 functional group.

Methylene - The CH_2 functional group.

Mid-infrared Radiation - Light between 4000 and 400 cm^{-1}.

Near-infrared Radiation - Light between 14,000 and 4000 cm^{-1}.

Nitrate - The NO_3 functional group.

Nitrile - The C≡N functional group.

Nitro - The NO_2 functional group.

Normal Modes - The constituent vibrations of any mechanical system. The complex vibrational motion of any molecule can be resolved into its normal mode vibrations.

Normalization - The process of taking a spectrum and dividing all the absorbances in a spectrum by the largest absorbance in the spectrum. This puts the Y-axis on a 0 to 1 scale. Used in *library searching* to remove the effects of spectra taken at different concentrations and pathlengths.

Olefins - See *alkenes*.

Ortho - An *isomer* of a disubstituted *benzene* ring, where the two substituents are on adjacent carbon atoms.

Overtone Transition - When *infrared absorption* causes a molecule's *vibrational quantum number* (V) to change from V=0 to V=2,3,4.... Overtone transitions give rise to infrared bands that are typically 10 to 100 times weaker than *fundamental transitions*.

Para - An *isomer* of a disubstituted *benzene* ring, where the two substituents are separated by two carbon atoms.

Phenol - An *aromatic alcohol*. A *benzene* ring with an O-H group bonded to it.

Phosphate - The PO_4 functional group.

Photon - A particle of light. Photons have energy, but have no mass.

Polymer - A molecule that consists of the same structural unit repeated over and over, like the links in a chain. These molecules typically have very high molecular weights.

Primary Alcohol - An alcohol where the *hydroxyl* group is attached to a *methylene* group. Alternatively, an alcohol where the *hydroxyl carbon* has one C-C bond.

Primary Amide - An *amide* where the nitrogen atom has one C-N bond and two N-H bonds.

Primary Amine - An *amine* that contains one C-N bond and two N-H bonds.

Quantum/Quanta - A packet(s) of energy.

Quantum Mechanics - The branch of physics that deals with quanta of energy, and describes the behavior of molecules, atoms, and sub-atomic particles.

Quaternary Carbon - A carbon atom bonded to four other carbon atoms. Alternatively, a carbon atom with four C-C bonds.

Reference Spectrum - In *spectral subtraction*, the spectrum of a pure material that is subtracted from the *sample spectrum*.

Ring Modes - Vibrations due to the stretching of the C-C bonds in a benzene ring.

Salt - A chemical formed by the reaction of an acid and a base.

Sample Spectrum - In *spectral subtraction*, the spectrum of a mixture from which the *reference spectrum* is subtracted.

Saturated - A *hydrocarbon* molecule that contains a maximum number of hydrogen atoms. All *alkanes* are saturated molecules.

Search Algorithm - A mathematical method used to compare spectra in *library searching* to measure the similarity of spectra. The application of a search algorithm results in the calculation of a *hit quality index*.

Secondary Alcohol - An *alcohol* where the *hydroxyl* group is attached to a *methine* carbon. Alternatively, an *alcohol* where the *hydroxyl carbon* has two carbon atoms bonded to it.

Secondary Amide - An *amide* that contains two C-N bonds and one N-H bond.

Secondary Amine - An *amine* that contains two C-N bonds and one N-H bond.

Silanol - The Si-OH group. Typically found on the surface of silica.

Silicones - The commercial name of *siloxanes*.

Siloxanes - *Polymers* that consist of Si-O-Si chains with pendant organic structural groups.

Spectral Subtraction - The process by which a *reference spectrum* is subtracted from a *sample spectrum*. Ideally, the bands due to the reference material are removed from the *sample spectrum*.

Subtraction Factor - In *spectral subtraction*, a number that is multiplied by the *reference spectrum* before the subtraction is performed.

Sulfate - The SO_4 functional group.

Sulfoxide - A functional group that contains a sulfur atom with two C-S bonds and one S=O bond.

t-Butyl - A *branch point* where three *methyl* groups are attached to a *quaternary carbon*.

Tertiary Alcohol - An *alcohol* where the *hydroxyl carbon* has three C-C bonds.

Tertiary Amide - An *amide* where the nitrogen atom contains three C-N bonds and no N-H bonds.

Tertiary Amine - An *amine* where the nitrogen atom contains three C-N bonds and no N-H bonds.

Thiol - The S-H functional group.

Trans - In disubstituted *alkenes*, an *isomer* where the two carbon-bearing substituents are on opposite sides of the C=C bond.

Transmittance - A unit used on the Y-axis in *infrared spectra*. It is measured as a percentage, and compares the intensity of light with no sample in the infrared beam to the intensity of light with a sample in the beam.

Unsaturated - A *hydrocarbon* molecule that contains less than the maximum number of hydrogen atoms it can have. All *alkenes* are unsaturated molecules.

Vibrational Interaction - This interaction occurs when two vibrations have the same symmetry and about the same energy. When one vibration is excited by *infrared absorption*, so is the second vibration. The energy of one vibration is raised, while the other is lowered. This causes the position of infrared bands due to these vibrations to appear in unexpected positions. Additionally, the intensities of the bands are different than would normally be expected.

Vibrational Quantum Number - Used to denote the vibrational energy levels of a molecule, starting with 0 at the lowest energy. The number is denoted with the letter V. Thus, a molecule at the V=1 level has one *quantum* of vibrational energy.

Vibrational Spectroscopy - The study of the interaction of light with matter when the interaction results in molecular vibrations.

Vinyl - A terminal *alkene*. The -CH=CH$_2$ group.

Wavelength - The distance between adjacent crests (or troughs) of a light wave.

Wavenumbers - The number of undulations a wave undergoes per centimeter, measured as a number per cm^{-1}. The most commonly used X-axis unit in *infrared spectra*.

Index

A

Absorbance, 5, 6–7
   Beer's Law, 6, 19, 196
   mechanics of, 7–14
   necessary conditions for, 14
   in spectral subtraction, 196
Absorptivity, 19
ABS rubber, 187
Acetamide, 125, 126
Acetate esters, 109, 232
Acetic acid, 109, 125
Acetone, 19, 92, 93, 94
Acetonitrile, 142
Acetophenone, 92, 93, 95
Acetylene, 47
Acetyl salicylic acid, 124, 235–236
Acid anhydrides, 105–108, 246
Acids. See Carboxylic acid salts; Carboxylic acids
Acrylates, 43, 184, 186–187
Acrylic acid, 184
Acrylonitrile, 187
Adsorbants, 169
Advantages and disadvantages, 1–2, 5
Alcohols, 67–72, 75, 83–85, 216–219, 244
Aldehydes, 18, 96–98, 114, 118–119, 228–229, 245
Aldehydic carbon, 96
Aldehydic hydrogen, 96, 97, 114
Aliphatic hydrocarbons, 31
Alkanes, 31–42, 243
   branched chain, 38–42, 243
   cyclic, 180
   straight chain, 32–38
   structure and nomenclature, 31–32
Alkenes, 28, 43–47, 210, 211
Alkyl aryl ether, 76, 78, 80
Alkyl ketone, 93, 96
Alkynes, 28, 47–48, 244
Alpha carbons, 92–93, 96, 98, 99, 108, 112, 126, 134, 141
Amide nitrogen, 125, 126
Amides, 125–132, 147–148, 237–238, 246–247
Amine nitrogen, 135
Amines, 135–141, 149–150, 239–240, 247
Amino acids, 132
Anharmonic vibration, 17–18
Anhydrides, acid, 105–108, 246
Aniline, 135, 149, 239
Anisole (alkyl aryl ether), 76, 78, 80
Answer key, 207–242
Area, of peak, 19
Aromatic acids, 99
Aromatic amines, 135, 138, 140, 141
Aromatic carbonates, 114
Aromatic esters, 108, 112, 115, 246
Aromatic ethers, 78
Aromatic hydrocarbons, 31, 49–58, 63. *See also individual compounds*
Aromatic nitriles, 143
Aromatic region, 25, 43, 70, 75, 92, 244
Artifacts, 27, 198–199, 200
Aryl/alkyl carbonates, 114
Aryl ether, 78
Aryl ketone, 93, 96
Aspirin, 124, 235–236
Asymmetrical vibrations. *See* Symmetrical and asymmetrical vibrations
Atlases, spectral, 195

B

"Background" spectrum, 6
"Ball and spring" model, 9, 15–18
Bandwidth. *See* Peak width
Baseline correction, 27
Beer's Law, 6, 19, 196
Bending vibrations, 35–36, 40, 45
Benzaldehyde, 118, 228
Benzamide, 128, 129
Benzene ring, 19, 21, 25, 28. *See also individual aromatic compounds*
   infrared spectrum, 50–52, 202, 203, 244
   in a nitro functional group, 144
   structure and nomenclature, 31, 49
   substituted, 49, 75
Benzenethiol, 154, 155
Benzoic acid, 100, 102, 112, 141
Benzonitrile, 21, 23, 141
Benzophenone (diphenylketone), 93
Benzyl acetate, 122, 232
Benzyl alcohol, 84, 217–218
Benzyl ether, 86, 220
Bond dipole moment, 9, 12
Bond distance, 12
Branch point, 38, 39, 78
Bromine, 163
Bromobenzene, 203
Butadiene, 187
t-Butanol, 85, 219
tert-Butylbenzene, 66, 215
t-Butyl group, 39, 40, 78, 81, 234
Butyl methyl ether, 88, 222

C

Calcium carbonate, 165, 172
Calcium sulfate, 166, 168
Carbonates
   inorganic, 171, 172, 249
   organic, 112–114, 115, 246
Carbon dioxide, 7, 10
Carbonyl group, 5, 91–115, 125–126, 133, 187
Carboxylates. *See* Carboxylic acid salts
Carboxylic acids, 98–102, 114, 120, 133, 230, 245

Carboxylic acid salts, 103–105, 115, 121, 130, 231, 245
Chemical environments, 20
Chemical structure, 1. *See also* Crystal structure; Functional groups
Chlorine, 163, 248
Chloroethane, 203
Chloroform, 162, 163, 201
Combination band, 14
Computer databases, of spectra, 195, 200
Concentration of sample
    Beer's Law and, 6, 19, 196
    peak intensity and, 1
    transmittance and, 7
Conjugation, 92, 93, 98, 112, 126
Crude oil, 31
Crystal structure, 165, 168–169, 180
Cyclohexane, 61, 180, 202, 203, 210
Cyclohexanol, 210

D

Dacron, 184
Deconvolution, 25–26
Delocalized electrons, 49
Detection limit, 1
Deuterium, 17
p-Diacetyl benzene, 117, 226–227
Dialkyl ether, 76
Diaryl ether, 78
Diaryl ketone, 93, 96
1,1-Dichloroethane, 203
Diethyl carbonate, 113
Diethyl ether, 76, 77
Diisobutyl phthalate, 123, 233–234
Diisocyanates, 187–190, 249
N,N-Dimethylacetamide, 126
N,N-Dimethyl amine, 141
N,N-Dimethylaniline, 135
Dimethyl benzene. *See* Xylene
Dimethylbutane, 39
2,3-Dimethylbutane, 40, 41
3,3-Dimethyl-1-butyne, 47, 48
Dimethyl ketone (acetone), 19, 92, 93, 94
N,N-Dimethylpropylamine, 135
Dimethyl sulfoxide (DMSO), 15, 156, 157
Diphenylketone, 93
Dipole moment, 8, 9–12
    in a carbonyl group, 91
    inorganic compounds, 165
    of the N-H bond, 127
    for symmetrical *vs.* asymmetrical stretch, 33, 76
Distillation, 25
DMSO (dimethyl sulfoxide), 15, 156, 157
Double bonds
    carbon-carbon, 43, 44, 45, 47. *See also* Carbonyl group
    carbon-nitrogen, 125, 187
    carbon-oxygen, 92, 128, 130, 133, 186, 187, 226
    sulfur-oxygen, 155, 156, 158, 168, 248

E

Economic aspects, 2
Electric vector, 2, 3, 7–10
Electromagnetic radiation, 2
Electronegativity, 9
Electronic effects, 17
Electron orbitals, 92, 103, 126, 143
Energy
    of molecules, 12–14
    of a photon, 9, 10, 13–14
    relationship with wavenumber, 4
Engineering plastics, 184–193
Equilibrium bond distance, 12
Esters, 108–112, 115, 183–184, 187, 232–234, 246
Ethanethiol, 153, 155
Ethanol, 67, 68–69, 70, 72, 109
Ether carbons, 75
Ethers, 75–81, 86–88, 220–222, 245
Ethyl acetate, 109, 110
Ethyl alcohol. *See* Ethanol
Ethylbenzene, 63, 212
N-Ethylformamide, 147, 237
Exercises
    answer key, 207–242
    problem spectra, hydrocarbons, 58–66
"Expert" programs, 204
Explosives, 143

F

Far-infrared radiation, 4
Fats, 43, 45, 108
Fermi resonance, 18, 96, 98, 100
Filtration, 25
Fingerprint region, 25
Fluorine, 163, 248
Fourier Transform Infrared Spectrometer (FTIR), 1, 200
Frequency, 3, 4
Functional groups, 1, 5. *See also* Methyl group; Methylene group
    of alkenes, 43–44
    t-butyl, 39, 40, 78, 81
    carbonyl, 91–115
    with the C-O bond, 67–81
    constancy of wavenumber, 21
    gem-dimethyl, 39, 40
    group wavenumbers for, 27–28, 243
    hydroxyl, 67, 72
    isopropyl, 39, 40, 70
    ketone, 93
    methine, 39, 67
    methoxy, 78, 81
    nitro, 143–146
    thiols, 153–155, 248
    vinyl, 43, 44, 45, 47
Fundamental transition, 14

G

gem-dimethyl group, 39, 40
Gemstones, 169
Ground vibrational state, 14
Group wavenumbers, 21, 25
Gypsum, 166, 168

H

Halogenated organic compounds, 160–163, 248
Halogens, 160, 248
Harmonic oscillator model, 15–17
Heat, 5
Hertz, 3
Hexane, 17, 20, 32, 33
    infrared spectrum, 34, 36, 38, 209
    umbrella mode, 35, 211
1-Hexene, 62, 211
High density polyethylene, 177, 179–180
Hit quality index (HQI), 25, 26–27, 199, 201–202
Homonuclear diatomic molecules, 2, 12
Hooke's Law, 15–16
Hydrocarbons, 31
    alkanes, 31–42
    alkenes, 43–47, 210, 211
    alkynes, 47–48
    aromatic, 31, 49–58
    estimation of chain length, 36, 38
Hydrogen bonds
    in alcohols, 68
    in amides, 127–128
    in amines, 135–136
    in the carbonyl functional group, 92
    in carboxylic acids, 99
    in silica compounds, 169
    in thiols, 155
Hydrogen chloride, 8–10, 140
Hydroquinone, 89, 223
p-Hydroxy anisole, 90, 224
Hydroxybenzene. *See* Phenols
Hydroxyl group, 67, 72, 99
p-Hydroxy phenol, 89, 223

I

Identities, 1, 26–27
Imide nitrogen, 132–133
Imides, 132–134, 190, 192, 247
Induced dipole moment, 10, 11
Infrared active/inactive, 10, 14
Infrared spectra, 1, 5–7
    acid anhydrides, 105–108
    alcohols, 68–72
    aldehydes, 96–98
    alkanes, 32–42
    alkynes, 47–48
    amides, 128–132
    amines, 136–141
    aromatic hydrocarbons, 50–58
    carboxylic acids, 99–100
    carboxylic acid salts, 103–104
    esters, 108–109, 110–111
    ethers, 76, 77, 79–80
    exercise samples, answer key to, 207–242
    exercise samples, hydrocarbons, 58–66
    imides, 133–134
    inorganic compounds, 165–167
    interpretation of, 25–28, 195–205
    ketones, 93–95
    manipulation of, 25–26
    nitriles, 141–143
    nitro group, 144–146
    organic carbonates, 114
    phenols, 73
    silica compounds, 169–170, 173
    siloxanes, 159, 160
    sulfur compounds, 157, 158, 168
    thiols, 154, 155
    water, 74, 75
Infrared spectrometer, 1
Inorganic carbonates, 171, 172, 249
Inorganic compounds, 165–174
Inorganic sulfates, 248
In-plane bending
    alcohols, 69, 70, 71
    amides, 128, 130, 237, 238
    amines, 138
    carboxylic acids, 100, 101, 102, 230
    esters, 235
    imides, 133
    inorganic carbonates, 171
    nitrates, 173, 174
    phenols, 72, 73, 223, 224
Intensity, peak, 1, 5, 18–19
Intermolecular interactions, 20, 22
Internet, use in infrared interpretation, 204–205
Interpretation of spectra, 25–29
    aids for, 195–205
    mixtures, 25–26, 196–199, 202
    performance of identities, 26–27
    10-step approach to, 27–28
Iodine, 163
Isobutanol, 219
Isobutyl, 39, 40, 234
Isomers, 39
    of amides, 127
    rotational, 160
    of substituted benzene rings, 49–50
Isooctane, 40, 42
Isopropyl alcohol, 70, 71, 72, 75
Isopropylbenzene, 64, 213
Isopropyl group, 39, 40, 70
Isovaleraldehyde, 97, 98

K

Kapton, 133, 190, 192
Keratin, 132
Ketones, 92–96, 114, 116–117, 225–227, 245
Kevlar, 148, 184, 238

L

Lexan, 114, 190, 191
Library searching, 25, 28, 199–203, 204–205
Light, 2–4, 13
Light boxes, 195
Low density polyethylene, 177–178, 180
Lucite, 187

M

Mass, of a molecule, 15–17
Mass effect, 17
Mass spectroscopy, 28
MEK (methyl ethyl ketone), 116, 225
Mercaptans. *See* 153–155
meta isomers, 50, 54
Metal atoms, 165
Methanol, 72, 112
Methine group, 39, 67, 180, 187, 244
Methoxy group, 78, 81
p-Methoxy phenol, 90, 224
N-Methyl acetamide, 126
Methylbenzene (toluene), 52, 53, 54
Methyl benzoate, 111
N-Methylbutylamine, 150, 240
2-Methylbutyric acid, 99, 100, 101
N-Methylcyclohexylamine, 138, 139, 140, 141
Methylene group, 32, 33, 35, 36, 38, 210, 244
  in acrylates, 187
  attached to alcohols, 67, 70
  attached to aldehydes, 98
  attached to an oxygen, in ethers, 78
  attached to ketones, 225
  in nylons, 184
  in polymers, 180
N-Methylethylamine, 135
Methyl ethyl ketone (MEK), 116, 225
Methyl group, 32, 33, 35, 244
  attached to alcohols, 70
  attached to amine nitrogen, 140–141
  attached to an oxygen, ethers, 78
  attached to benzene rings, 54, 58, 244
  attached to ketones, 225, 226
  attached to silicon atom, 160
N-Methylphthalimide, 133
Methyl *t*-butyl ether (MTBE), 78, 79
Microwaves, 4
Mid-infrared radiation, 4
Mixed carbonates, 114
Mixed ethers, 76, 78
Mixed ketones, 93
Mixtures, 2, 25–26, 27, 196–199, 202
Monatomic ions, 2
Monomers, 177
MTBE (methyl *t*-butyl ether), 78, 79
Mylar, 184

N

Nearest neighbor interactions, 20, 22
Near-infrared radiation, 4, 14
Newton's second law, 16
Nitrates, 171, 173, 174, 249
Nitrile carbon, 141
Nitriles, 141–143, 151, 241, 247
Nitrogen
  effects of, 35
  electronegativity, 125, 127
  in nitrates, 171, 173, 174
  in organic nitrogen compounds, 125–152, 247
  in polyurethanes, 187
Nitro group, 143–146, 152, 242, 248
Nitro nitrogen, 143
m-Nitrotoluene, 145
O-Nitrotoluene, 152, 242
Noble gases, 2
Nomex, 184
Normalized spectra, 200
Normal modes, 7
Nuclear magnetic resonance (NMR), 28, 144, 242
Nylon-3,5, 184
Nylon-6,6, 128, 130, 131, 184, 185
Nylons, 128, 184

O

Octadecane, 65, 214
Octane, 17, 60, 78, 209
Octanoic acid, 120, 230
1-Octene, 43, 44, 45, 46
Octylaldehyde, 119, 229
Odor, of thiols, 155
Oils, 43
Olefins. *See* Alkenes
Organic carbonates, 112–114, 115, 246
Organic compounds, halogenated, 160–163
Organic nitrogen compounds, 125–152
Organic silicon compounds, 158–160
Organic sulfur compounds, 153–158
ortho isomers, 50, 54
Out-of-plane bending
  alcohols, 70, 71, 216, 217, 219
  aldehydes, 228
  amides, 130, 237, 238
  amines, 137, 138, 239
  aromatic hydrocarbons, 51, 52, 53, 54, 55, 57, 65
  carboxylic acids, 100, 101, 102, 230
  esters, 232, 233, 235
  ethers, 220, 221
  examples, 210, 211, 212, 213, 215
  inorganic carbonates, 171, 172
  ketones, 226
  nitrates, 173, 174
  nitriles, 241
  nitro group, 144, 242
  phenols, 75, 223, 224
  polymers, 182

Ovelap between peaks, 21, 25
Overtone transition, 14
Oxygen
  in the carbonyl group, 91
  effects of, 35
  in esters, 108–109
  in hydrogen bonding, 68, 166, 245
  in methoxy group, 78, 81

P

para isomers, 50, 54
Peak area, 19
Peak intensity, 1, 5, 18–19
Peak position, 1, 15–18
Peaks, overlap between, 21, 25
Peak shape, 1
Peak width, 1, 19–21, 68
Petroleum jelly, 36, 37, 38, 180, 209
Petroleum products, 31
Phenols, 67–68, 72–75, 89–90, 223–224
Phenyl ether, 87, 221
Phenyl methyl ether (alkyl aryl ether), 76, 78, 80
Phosphates, 173, 249
Photons, 4
  energy of, 9, 10, 13–14
  transmittance and absorbtion by water, 12
Phthalic acid, 133
Phthalic anhydride, 105, 106
Phthalimide, 133, 134
Planck's Constant, 13
Plasticizers for polymers, 234
Plastics, 133
  engineering, 184–193
  recyclable, 177–184
Plexiglass, 187
PMMA (polymethyl methacrylate), 186, 187
Polyacrylonitrile, 187
Polyamides, 184–185
Polycarbonates, 21, 24, 190, 191, 197, 198
Polydimethylsiloxane, 158–159, 160
Polyether foam, 188, 189
Polyethylene, 59, 177–180, 208
Polyethylene terephthalate (PET), 183, 184
Polyimides, 190, 192
Polymers, 43, 177–193
Polymethacrylic acid, 187
Polymethyl methacrylate (PMMA), 186, 187
Polyparaphenylene terephthalamide, 148, 238
Polypropylene, 43, 180, 181
Polystyrene, 24, 43, 180, 182, 197, 198
  peak intensity, 18–19
  peak width, 21
  transmittance spectrum, 6
Polytetrafluoroethylene, 161, 163, 190, 193
Polyurethanes, 187–190
Polyvinylacetate, 43
Polyvinylchloride, 43
Primary amides, 126, 246
Primary amines, 135, 247
Process of elimination, 26

n-Propanol, 83, 216
Propionamide, 126
Propylamine, 136, 137
Proteins, 128, 132
Purification methods, 25

Q

Quantum mechanics, 12
Quartz, 168–169
Quaternary carbon, 39

R

Radio waves, 4
Raman spectroscopy, 28
Recrystallization, 25
Recyclable plastics, 177–184
Reduced mass, 15, 16, 18
Reference spectrum, 25–26, 196–199
Repeat unit, 177, 180
Resolution, 27, 82
Resources, for spectral libraries, 200, 204–205
"Ring-bending" band, 52, 53, 54, 56
  examples, 182, 212, 213, 215, 217, 220, 221
Ring modes, 52, 53, 55, 56, 57, 212, 213, 215
Rocking vibration
  aldehydes, 229
  alkanes, 34, 36, 37, 38, 208, 214
  polymers, 178–179, 180
  siloxanes, 159, 160
Rotamers, 160
Rubber, 187
Rubbing alcohol, 70, 71, 72, 75
"Rule of Three," 108–109, 112, 184, 190

S

Salts. *See* Carboxylic acid salts
Samples, 25, 27, 196
Saturated alcohols, 67, 70–72
Saturated aldehydes, 98
Saturated amines, 135, 138, 140, 141
Saturated carbonates, 114
Saturated esters, 108, 109–112, 115, 235, 246
Saturated ethers, 76, 78, 223, 224
Saturated hydrocarbons, 31, 209
Saturated ketones, 93
Saturated nitriles, 141, 142, 143
"Scissors" vibration
  alkanes, 36, 208
  amides, 128, 129
  amines, 136, 137, 138, 239
  nitro group, 144, 145, 146, 242
  polymers, 180
  water, 75, 208
Search algorithms, 201
Search report, 201
Secondary alcohols, 67, 70
Secondary amides, 126, 128, 130, 131, 138–140, 247
Secondary amines, 135, 247

Shape of peak, 1
Signal-to-noise ratio (SNR), 27, 200
Silanol bonds, 169, 249
Silica, 170, 249
Silica compounds
   inorganic, 168–170, 173, 249
   organic, 158–160
   uses for, 169
Silicates, 169, 173, 248
Siloxanes, 158–160, 248
Smoothing, 27
Sodium acetate, 103
Sodium benzoate, 121, 231
Sodium hydroxide, 103
Sodium nitrate, 174
Sodium stearate, 103
Sodium sulfate, 167, 168
Software packages, 204
Solvents, 155
Spectral atlases, 195
Spectral library searching, 25, 28, 199–203, 204–205
Spectral subtraction, 25, 27, 196–199
Stearic acid, 103
Stretching vibrations, 32–34, 35–36, 43–44
Styrene, 187
Substituted alkenes, 43, 44, 47
Substituted alkynes, 47
Substituted benzene rings, 49, 52–58, 92, 244
"Subtract and search again," 25, 196–199, 202
Subtraction, spectral, 25, 27, 196–199
Subtraction artifacts, 27, 198–199, 200
Subtraction factor, 196–197
Subtrahend, 25–26, 196
Succinic anhydride, 105, 106
Sulfates, 156, 158, 168, 173, 248
Sulfite, 156, 248
Sulfonate, 156, 158, 248
Sulfone, 156, 158, 248
Sulfoxide, 155, 248
Sulfur compounds
   inorganic, 167, 168, 248
   organic, 153–158, 248
Symmetrical and asymmetrical substitution, 47, 54
Symmetrical and asymmetrical vibrations
   acid anhydrides, 106, 107, 108
   alcohols, 69, 70, 71, 72, 216, 217, 219
   aldehydes, 229
   amides, 237
   amines, 136, 137, 138, 139, 140, 141, 239, 240
   branched alkanes, 40, 41, 42
   carboxylic acids, 230
   carboxylic acid salts, 103, 104
   ethers, 76, 77, 78, 79, 80, 81, 220, 222
   examples, 208, 209, 210, 211, 213, 214, 215
   nitro group, 144, 145, 146, 242
   phenols, 73, 75
   polymers, 178–179, 181, 182
   polyurethanes, 190
   siloxanes, 159, 160
   straight-chain alkanes, 32–33, 35–36, 37, 38

T

Teflon, 161, 163, 190, 193
tert-butylbenzene, 66, 215
Tertiary alcohols, 67, 72, 75, 219
Tertiary amides, 126, 130, 247
Tertiary amines, 135, 140
Thiol group, 153–155, 248
Toluene, 52, 53, 54
Transmittance, 6–7, 196
Trinitrotoluene, 143
Triple bonds
   carbon-carbon, 47, 48
   carbon-nitrogen, 125, 141–143, 241, 247

U

Ultraviolet light, 4
Ultraviolet/visible light spectroscopy, 28
Umbrella bending mode
   alcohols, 70, 216, 219
   aldehydes, 229
   alkanes, 34, 35–36, 37, 38, 40
   amides, 237
   amines, 240
   carboxylic acids, 100, 230
   esters, 232, 233, 235
   ethers, 78, 81, 222
   examples, 211, 212, 213, 214, 215
   ketones, 225, 226
   nitro group, 242
   phenols, 75
   polymers, 178–179, 180, 181
   siloxanes, 159, 160
Unsaturated alcohols, 67
Unsaturated hydrocarbons, 31
Unsaturated phenols, 223
Urethanes, 187, 249

V

Valeric anhydride, 106
Vibration, symmetrical vs. asymmetrical, 32–33, 35–36, 37, 38, 40
Vibrational interaction, 18
Vibrational quantum number, 12, 14
Vibrational spectroscopy, 7
Vinyl group, 43, 44, 45, 47, 187, 210, 244
Vinylidine, 43, 44, 45, 47, 210, 244

W

"Wag" band, 47, 48, 128, 138, 139, 140, 160, 240
Water, 2
   absorbance of, 14
   energy of, 12–13
   infrared spectrum, liquid, 74, 75
   in inorganic compounds, 165
   intermolecular interactions, 20, 22
   normal modes of, 7

Waters of hydration, 165, 168
Wavelength, 2–4
Wavenumber, 3–4
    carbon double bonds, 45, 91, 100, 112, 187, 243
    carbon-nitrogen bonds, 128, 130, 187, 243
    carbon triple bonds, 47, 48, 243
    group, origin of, 21, 25
    halogens, 160, 248
    relationship with peak intensity, 19
    relationship with radiation intensity, 5
    of ring modes, 52, 53, 55, 56, 57
    silanol bonds, 169, 249
    sulfur bonds, 156, 158, 248
    tables, 243–249

Width of peak, 1, 19–21, 68
World Wide Web, 204–205

X

X-rays, 4
Xylene, 50, 54
    meta-, 56
    ortho-, 55
    para-, 57
    substituted, 54–57

Z

Zinc hydroxide, 103
Zinc stearate, 103, 104